Minerals, Rocks and Inorganic Materials

Monograph Series of Theoretical and Experimental Studies

3

Edited by

W. von Engelhardt, Tübingen · T. Hahn, Aachen

R. Roy, University Park, Pa.

J. W. Winchester, Tallahassee, Fla. · P. J. Wyllie, Chicago, Ill.

Subseries:
Isotopes in Geology

Bruce R. Doe

Lead Isotopes

With 24 Figures

Springer-Verlag New York · Heidelberg · Berlin 1970

Dr. *Bruce R. Doe*
U.S. Geological Survey
Denver, Colorado/U.S.A.

Present Address:
National Aeronautics and Space Administration Code MAL
Washington, D.C./U.S.A.

Publication authorized by the Director,
U.S. Geological Survey

ISBN 978-3-642-87282-2   ISBN 978-3-642-87280-8 (eBook)
DOI 10.1007/978-3-642-87280-8

Title-No. 3733

Foreword

This series of monographs represents continuation on an international basis of the previous series MINERALOGIE UND PETROGRAPHIE IN EINZELDARSTELLUNGEN, published by Springer-Verlag. The voluminous results arising from recent progress in pure and applied research increase the need for authoritative reviews but the standard scientific journals are unable to provide the space for them. By their very nature, text-books are unable to consider specific topics in depth and recent research methods and results often receive only cursory treatment. Advanced reference volumes are usually too detailed except for experts in the field. It is often very expensive to purchase a symposium volume or an "Advances in..." volume for the sake of a specific review chapter surrounded by unrelated chapters. We hope that this monograph series will by-pass these problems in fulfilling the need. The purpose of the series is to publish reviews and reports of carefully selected topics written by carefully selected authors, who are both good writers and experts in their scientific field. In general, the monographs will be concerned with the most recent research methods and results. The editors hope that the monographs will serve several functions, acting as supplements to existing text-books, guiding research workers, and providing the basis for advanced seminars.

July 1970

W. VON ENGELHARDT, Tübingen
T. HAHN, Aachen
R. ROY, University Park, Pa.
J. W. WINCHESTER, Tallahassee, Fla.
P. J. WYLLIE, Chicago, Ill.

Preface

Sub-series on "Isotopes in Geology". The branch of earth sciences dealing with the application of isotope studies to the interpretation of earth processes has provided much information in recent years, very little of which has made its way into text-books. Teachers and research workers therefore have to dig assiduously through widely scattered journals if they to follow developments. In the Preface of his 1963 book on "Progress in Isotope Geology", K. RANKAMA concluded: "isotope geology now covers such a vast and rapidly growing field" that "future reviews of research will require the time and effort of more than one individual". For these reasons, I have planned a series of monographs reviewing and explaining basic theory, methods of measurements, the results that have been obtained in recent years, and their applications to petrology and problems in the earth sciences. This first volume is a detailed review of the variation of lead isotopes in minerals and rocks, with explanations and applications. The volume is slender, but author BRUCE DOE has packed into it a vast amount of data.

July 1970

PETER J. WYLLIE
University of Chicago

Contents

I. Introduction

More than 1,000 papers have been written concerning lead isotopes (Doe, 1968b) since Aston's first abundance measurements were made by the mass spectrograph (Aston, 1927, 1929). Aston's first measurements were exciting as they resulted in the discovery of ^{207}Pb by Aston and discovery of ^{235}U, estimation of its half-life within better than a factor of two, and the estimation of the age of the earth within a factor of two by Rutherford (1929). Precision measurements by mass spectrometric methods (Nier, 1938, 1939; Nier et al., 1941) introduced modern lead isotope studies. The isotopic composition of lead varies because of the radioactive decay of ^{238}U to ^{206}Pb, ^{235}U to ^{207}Pb, and ^{232}Th to ^{208}Pb (see Appendix A for decay chains). One isotope, ^{204}Pb, has no long lived radioactive parent. Thus the process producing isotopic differences in lead isotopes should not be confused with the physico-chemical fractionation processes that cause isotopic differences in stable isotopes of light elements such as carbon, oxygen and sulfur. The prime factors controlling partitioning of isotopes are the mass separation divided by the atomic weight and changes in the oxidation state. Sulfur, for example, has a complex geochemistry characterized by many oxidation states and a range of ^{32}S/^{34}S in nature of about 15 percent, whereas silica, with approximately the same percent mass range for ^{28}Si/^{30}Si but with a very simple geochemistry, has less than 1/10 the range found for sulfur isotopes.

For lead, the greatest *equilibrium* fractionation that would be expected would be 0.05 percent of ^{208}Pb/^{204}Pb (see review in Russel and Farquhar, 1960). The greatest *single stage disequilibrium* fractionation that would be expected would be molecular distillation (preferential volatilization of the lighter isotope into a vacuum relative to the heavier isotope). For this mechanism, the vapor may be enriched by $\leq \sqrt{208/204}$ or about 1 percent *maximum*. Some confusion has arisen here with regard to the separation of lead isotopes by physico-chemical methods as ^{235}U has been greatly enriched relative to ^{238}U synthetically by using many stages of disequilibrium physico-chemical processes. It should be remembered, however, that the variation so far observed for ^{238}U/^{235}U in nature is only 0.07 percent of the ratio (for one review of natural variations in uranium isotopes see Doe and Newell, 1966). Other empirical and theoretical justification for the lack of significant lead isotope effects due to physico-chemical processes in nature are reviewed in DOE *et al.* (1967). They point out that studies of rubidium, chlorine, the non-radiogenic isotopes of strontium, and silver (an element that is geochemically rather similar to lead) have failed to find any isotopic variations greater than 0.1 percent. They also point out that isotopic variations caused by physico-chemical processes would be proportional among the lead isotopes, i.e. the percentage effect on $^{206}Pb/^{204}Pb$ would be half that observed for $^{208}Pb/^{204}Pb$ and that for $^{207}Pb/^{204}Pb$ would be 3/4 that observed for $^{208}Pb/^{204}Pb$. While such variations do occur in the instruments used to measure isotopic abundances, there is no evidence of these sorts of systematics in the lead isotope variations in nature. Therefore, consideration of both the theory of physico-chemical induced isotope effects plus the available empirical data indicate that natural equivalents of thermal diffusion and adsorption columns do not significantly affect the isotope ratios of the heavier elements such as lead.

Additional reviews for the reader who is interested in more detailed discussions are those of RUSSELL and FARQUHAR (1960), HAMILTON (1965), TUGARINOV and VOYTKEVICH (1966), DARNLEY (1964); the series of papers by CATANZARO (1968), HART *et al.* (1968), and KANASEWICH (1968a) in the book "Radiometric Dating for Geologists"; and the reports of GAST (1967), ZARTMAN (1969), and WASSERBURG (1966).

Lead isotope data will be discussed in three subsections of the monograph: II, U-Th-Pb dating; III, Common lead; and IV, Radioactive lead isotopes.

Acknowledgments

I particularly wish to thank IVAN MITTIN, NORMAN HUBBARD, and PRIESTLEY TOULMIN for translating several papers from the Russian to the English and LOUIS NICOLAYSEN, MARC GRÜNENFELDER, VIKTOR KOEPPEL, M. TATSUMOTO, JOHN ROSHOLT, and JOHN STACEY for spirited discussions on the manuscript. It is a pleasure to acknowledge that a part of this manuscript was completed while the author was an academic guest at the Institut für Kristallographie und Petrographie, Eidgenössische Technische Hochschule, in Zürich, Switzerland.

II. U-Th-Pb Dating

General. Three independent ages may be obtained in the U-Th-Pb system: $^{206}Pb/^{238}U$, $^{207}Pb/^{235}U$ or $^{207}Pb/^{206}Pb$, and $^{208}Pb/^{232}Th$. Emphasis has been placed on U-Pb dating because the value $^{238}U/^{235}U$ is a physical constant that permits internal treatment of the data not found in any other dating system. This treatment helps to eliminate the assumption that the phase being dated has remained closed to changes in the parent-daughter system. The theoretical systematics are expressed in Fig. 1. WETHERILL (1956a, b) showed that a phase,

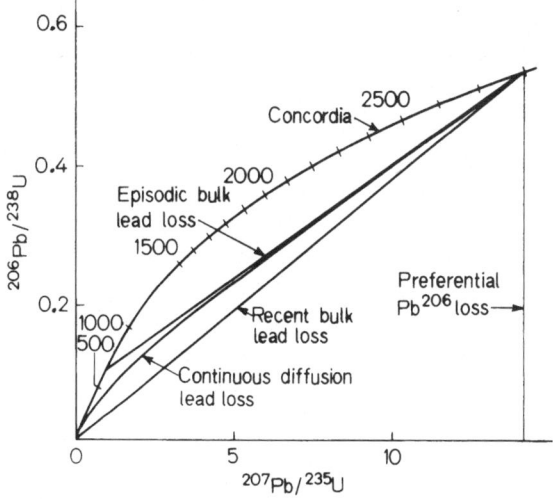

Fig. 1. Systematics of U-Pb dating (after CATANZARO and KULP, 1964)

which is subject to no lead loss or uranium gain (a closed system), will have $^{207}Pb/^{235}U$ ages equal to $^{206}Pb/^{238}U$ ages and that the data will lie along a curved line called *concordia*. In addition, he showed that phases subject to lead loss or uranium gain during a period of time that is short compared with the age of the phase (episodic bulk daughter losses or parent gains), recently or in the distant past, will have data that will lie along a straight line called *discordia*. The lower intersection of discordia with concordia represents the time of the episodic event and the upper intersection represents the age of the phase. NICOLAYSEN (1957) suggested that diffusion of lead out of a phase might

1*

take place at a constant rate over the entire history of the phase (continous diffusion), and TILTON (1960) showed that continuous diffusion with bulk lead loss or uranium gain also closely approaches a straight line. WASSERBURG (1963) considered the possible effects of a diffusion constant that varies with time, such as might be caused by radiation damage; and WETHERILL (1963) treated most other aspects of the theory of U-Pb behavior, such as continuous diffusion with superimposed episodic loss and the theory of uranium loss and bulk radiogenic daughter gain. The U-Th-Pb system has intermediate radioactive daughters, including the gas radon (see Appendix A). If there is radon loss over a long period of time, the calculated ages may be affected. If the system is subject to radon leakage, the result is reflected in preferential loss of ^{206}Pb, because the radon in the ^{238}U chain has a much longer half-life (3.8 days) than that in the ^{235}U chain (3.9 seconds). The longer half-life allows a longer time for radon to escape.

STIEFF and STERN (1961) and STIEFF et al. (1963) considered plots of data similar to the concordia plot wherein the common lead was not first removed in the age calculation. These plots allow calculation of the initial ^{207}Pb/^{206}Pb. Also valuable are plots of ^{206}Pb/^{204}Pb versus ^{238}U/^{204}Pb and ^{207}Pb/^{204}Pb versus ^{235}U/^{204}Pb, similar to the strontium isochron plot (or Bernard Price Institute plots), which also allow the initial values of ^{206}Pb/^{204}Pb and ^{207}Pb/^{204}Pb to be determined. These plots are best suited to situations in which there is not great radiogenic enrichment. One good example of this application is given by MIRKINA and MAKAROCHKIN (1966).

Though considerable data have been obtained on Th-Pb dating, this dating has received less emphasis because of the lack of a companion radioactive parent isotope to go with ^{232}Th. The theoretical behavior is, however, given by STEIGER and WASSERBURG (1966).

Most U-Th-Pb dating has been done on uraninite, pitchblende, and other uranium minerals. Uraninite under some conditions gives the most concordant ages but it is not a common accessory mineral in most rock types. Emphasis, therefore, will herein be placed on the more common accessories – zircon, sphene, apatite, members of the epidote-family, monazite – and on whole-rock dating. Zircon has been subjected to the most thorough and rigorous laboratory and field investigation. It is therefore chosen for first discussion and will be the basis for judging the behavior of the other more common accessories, in the order given above but with particular emphasis on sphene and apatite. Some other minerals will be briefly discussed: pyrochlore and other members of the columbite-tantalite family, xenotime, and glauconite.

A table of equations and parameters used in age calculations is given in Table 1. The decay constants are those preferred from the evaluation by ALDRICH and WETHERILL (1958) and also for ^{235}U by BANKS and SILVER (1966). Table 2 contains some satisfactory analytical techniques. The list is not comprehensive. Table 3 presents results on some of the better analyzed lead isotope standards.

Table 1. *Equations and parameters used in age calculations*

1. Basic equation: $-\dfrac{dN}{dt} = \lambda N$

   N = number of radioactive atoms
   t = time
   λ = constant of proportionality (decay constant)

2. Working equations (atomic ratios):

   a) $^{206}Pb/^{238}U$ age

   $$\left(\frac{^{206}Pb}{^{204}Pb}\right)_{observed} - \left(\frac{^{206}Pb}{^{204}Pb}\right)_{initial} = \left(\frac{^{238}U}{^{204}Pb}\right)_{observed} \cdot (e^{\lambda_8 T} - 1)$$

   b) $^{207}Pb/^{235}U$ age

   $$\left(\frac{^{207}Pb}{^{204}Pb}\right)_{observed} - \left(\frac{^{207}Pb}{^{204}Pb}\right)_{initial} = \left(\frac{^{235}U}{^{204}Pb}\right)_{observed} \cdot (e^{\lambda_5 T} - 1)$$

   c) $^{208}Pb/^{232}Th$ age

   $$\left(\frac{^{208}Pb}{^{204}Pb}\right)_{observed} - \left(\frac{^{208}Pb}{^{204}Pb}\right)_{initial} = \left(\frac{^{232}Th}{^{204}Pb}\right)_{observed} \cdot (e^{\lambda_2 T} - 1)$$

   d) *Pb-Pb age or isochron age*

   $$\frac{\left(\frac{^{207}Pb}{^{204}Pb}\right)_{observed} - \left(\frac{^{207}Pb}{^{204}Pb}\right)_{initial}}{\left(\frac{^{206}Pb}{^{204}Pb}\right)_{observed} - \left(\frac{^{206}Pb}{^{204}Pb}\right)_{initial}} = \left(\frac{^{235}U}{^{238}U}\right)_{observed} \cdot \frac{(e^{\lambda_5 T} - 1)}{(e^{\lambda_8 T} - 1)}$$

   $$\left(\frac{^{235}U}{^{238}U}\right)_{observed} = \frac{1}{137.8} \quad \text{(INGHRAM, 1946)}$$

   e) *Explanation of abbreviated symbols*

   T = geologic age
   λ_8 = decay constant for $^{238}U = 1.54 \times 10^{-10}$ yr^{-1} (KOVARIK and ADAMS, 1955)
   λ_5 = decay constant for $^{235}U = 9.71 \times 10^{-10}$ yr^{-1} (FLEMING et al., 1952)
   λ_2 = decay constant for $^{232}Th = 4.99 \times 10^{-11}$ yr^{-1} (KOVARIK and ADAMS, 1938; PICCIOTTO and WILGAIN, 1956)

   Initial ratio is that ratio which the phase contained when it was formed.

3. *Isotopic masses* (BHANOT et al., 1960)

$^{204}Pb = 204.038$	$^{235}U = 235.119$
$^{206}Pb = 206.040$	$^{238}U = 238.126$
$^{207}Pb = 207.042$	$^{232}Th = 232.111$
$^{208}Pb = 208.043$	

Table 2. *Some satisfactory analytical techniques for lead isotope studies*

Method	Minimum applicable content	Application and comment	Reference
Dissolution procedure			
Borax fusion[a] KHF fusion	10.0 ppm	General, especially on zircons and zircon-rich rocks	TILTON et al. (1955) CATANZARO and KULP (1964)
HF–HClO$_4$ HF–COOH$_2$	2.0 ppm	General, on nonrefractory materials (feldspars, monazite)	TILTON et al. (1955) SOBOTOVICH et al. (1963b)
Volatilization in inert atmosphere	0.5 ppm	On isotopically homogeneous materials (for lead isotopic composition only)	STARIK et al. (1958b) STARIK et al. (1958c)
Volatilization in vacuum			MASUDA (1964)

[a] Commercial borax of adequate purity in some 5-pound bottles can now be obtained. Tests run for this paper give the following results by isotope dilution (the lots quoted are no longer available):

Brand	Lot number	Pb content (hydrous borax)	Source of data
Union Chimique Belge	Not known	0.09, 0.05 (via ^{206}Pb)	PAUL PASTEELS (written commun., 1969)
FISHER	754996	0.096 ppm	This study
BAKER	32215	0.033 ppm via ^{206}Pb	This study
BAKER	34153	0.093 ppm	This study
BAKER	30154	0.011 ppm (via ^{206}Pb) 0.045 ppm (via ^{208}Pb)	MARC GRÜNENFELDER (written commun., 1968)
BAKER	37574	0.0067 ppm (via ^{206}Pb) 0.027 ppm (via ^{208}Pb)	GEORGE TILTON (written commun., 1968)

Table 2 (continued)

Method	Minimum applicable content	Application and comment	Reference
Volatilization in vacuum by melting	0.1 ppm	General; does not require sample to be pulverized (for lead isotopic composition only)	TATSUMOTO (1966b)
6N HCl 7N HNO$_3$	0.1 ppm	General, on acid-soluble minerals and rocks (carbonates, phosphates, and pyrite)	Many textbooks
Lead purification			
Double dithizone solution extraction	1 µg	In the absence of ^{209}Bi and ^{203}Tl and ^{205}Tl (zircons, feldspars)	TILTON et al. (1955)
Double chloride anion resin columns	5 µg	For zinc- and silver-poor materials (zircon and feldspars)	CATANZARO and GAST (1960)
Bromine anion resin columns	5 µg	For iron-rich materials (meteorites, basalts)	GAST et al. (1964), STRELOW and TOERIEN (1966)
Chloride anion resin column plus dithizone solution extraction	5 µg	For rocks of low or moderate iron content; most efficient to separate lead from ^{209}Bi, ^{203}Tl, ^{205}Tl and silver	DOE et al. (1967)
Barium coprecipitation plus dithizone solution extraction	5 µg	In the absence of ^{209}Bi and ^{203}Tl and ^{205}Tl (basalts, feldspars, zircons)	TATSUMOTO (1966a)
Electrodeposition onto platinum electrodes	500 µg	General, a very promising recent technique	SHIELDS (1967)
Electrodeposition onto filament material in 1N HCl with hydroxylamine	Not specified	General, a very promising recent technique	ALLEGRE et al. (1968)

Table 2 (continued)

Method	Minimum applicable content	Application and comment	Reference
Uranium and Thorium purification			
Solution extraction with methyl isobutyl ketone with salting out agents	0.01 µg U 0.1 µg Th	For purification of combined uranium and thorium	Tilton et al. (1955)
Anion exchange in nitrate form with HNO_3. A more quantitative method for thorium than the hexone procedure given above		For purification of combined uranium thorium	Tatsumoto (1966a)
Anion exchange in the chloride form	0.01 µg	For purification of uranium only on iron-free materials	Catanzaro and Kulp (1964)
Determination of concentrations			
Isotope dilution with purified [206]Pb for unradiogenic leads and [208]Pb for radiogenic leads [b]	0.01 µg	Lead	Tilton et al. (1955)
Isotope dilution with purified [235]U and [230]Th [c]	0.01 µg U 0.01 µg Th	Uranium and thorium	Rosholt et al. (1966)

[b] Purified [206]Pb and [208]Pb may be obtained from the Isotope Division, Oak Ridge National Laboratory, Union Carbide Nuclear Company, Post Office Box X, Oak Ridge, Tennessee 37830.

[c] Purified [235]U and [230]Th may be obtained on loan from Research Materials Coordinator, Chemistry Programs, Division of Research, U.S. Atomic Energy Commission, Washington, D. C. 20545. License and Loan Agreement must be obtained. The compounds as received from Oak Ridge National Laboratory are found to be close to stoichiometric. Calibration of uranium can be checked against NBS 950A (U_3O_8 standard) available from the U.S. Bureau of Standards, Washington, D.C., 20234, and thorium can be checked against Lindsay ThO_2 (code 116) available from Lindsay Chemical Division, American Potash and Chemical Corp., West Chigaco, Illinois. ThO_2 may be taken into solution by heating in a large excess of 6.2 N-HCl for a week or so at just below the boiling point.

Table 2 (continued)

Method	Minimum applicable content	Application and comment	Reference
Isotope abundance measurement			
Surface emission mass spectrometry[d]	10 µg	PbS in NH$_4$OH solution	TILTON et al. (1955); DOE et al. (1967)
		Lead oxalate in NH$_4$OH	COOPER and RICHARDS (1966a)
	0.001 µg	Lead in silica gels	AKISHIN et al. (1957); CAMERON et al. (1969)
	20 µg	Double spike technique	COMPSTON and OVERSBY (1969)
	100 µg	Crystals in H$_2$PO$_4$	KOSZTOLANYI (1965)
Thermal emission mass spectrometry[e]	500 µg	Pb(NO$_3$)$_2$ with NH$_4$OH	CATANZARO (1967, 1968); CATANZARO et al. (1968)
	100 µg	Uranium in HNO$_3$	SHIELDS (1966)
	5 µg	Uranium and thorium in HNO$_3$	ROSHOLT et al. (1966)
Electron bombardement mass spectrometry	200 µg	Lead tetramethyl	COLLINS et al. (1951); KOLLAR et al. (1960); ULRYCH and RUSSELL (1964)
	10 µg	Lead iodide	NIER (1938), DELEVAUX (1963); DOE et al. (1967)
Activation analysis (for ultralow lead contents)	0.01 µg	Neutron activation (^{208}Pb/^{204}Pb)	REED et al. (1960)
	Not stated	Alpha particle activation (^{206}Pb/^{204}Pb and ^{208}Pb/^{204}Pb)	COBB (1964)

[d] Analytical uncertainties are about 0.1 percent per mass unit separation of the measured ratio at one standard deviation without sample size control and about 0.05 percent per mass unit separation of the measured ratio at one standard deviation with sample size control (DOE et al., 1967).

[e] Analytical uncertainties are about 0.025 percent of each ratio measured independent of the mass unit separation at one standard deviation.

Table 3. *Absolute isotopic ratios of some standard samples*

	$^{206}Pb/^{204}Pb$	$^{207}Pb/^{204}Pb$	$^{208}Pb/^{204}Pb$	References
CIT lead standard[a]	16.625	15.475	36.300	CATANZARO(1967)
	16.62	15.49	36.34	DOE (1962b)
GS 4 lead standard[b]	16.158	15.406	35.841	CATANZARO (1967a)
NBS standard samples[c]:				
Common lead (SRM 981)	16.937	15.491	36.721	CATANZARO et al. (1968)
	$^{204}Pb/^{206}Pb$	$^{207}Pb/^{206}Pb$	$^{208}Pb/^{206}Pb$	
Common lead (SRM 981)	0.059042 ± 0.000037	0.91464 ± 0.00033	2.1681 ± 0.0008	CATANZARO et al. (1968)
Equal atom (SRM 982)	0.027219 ± 0.000027	0.46707 ± 0.00020	1.00016 ± 0.00036	CATANZARO et al. (1968)
Radiogenic lead (SRM 983)	0.000371 ± 0.000020	0.071201 ± 0.000040	0.013619 ± 0.000024	CATANZARO et al. (1968)

[a] A lead sample distributed by the geochemistry group at the California Institute of Technology.

[b] A lead sample distributed by the Isotope Geology Branch of the U.S. Geological Survey.

[c] Sets of the NBS standard samples may be ordered from Office of Standard Reference Materials, National Bureau of Standards, Washington, D.C. 20234. The order should stipulate: 1 g each of NBS SRM Nos. 981, 982, and 983 (lead reference standards). The price is 100 U.S. dollars for the set.

1. Zircon (ZrSiO$_4$)

Interest in the isotopic dating of zircon came early; the cyrtolitic variety of zircon was analyzed by NIER (1939). Interest was sustained by the chemical dating method (Pb-α or "Larsen method") (LARSEN et al., 1952). One valuable feature of zircon is that the content of common lead is small, and the resulting calculated age is insensitive to the isotopic composition of the contained common lead. With proper analytical techniques (STERN and ROSE, 1961; ROSE and STERN, 1960), chemical dating of zircons appears to give an age approximating the ^{207}Pb/^{235}U isotopic age. As the sample size of zircon required for analysis is small, chemical dating may be still found useful in outlining problems.

The first isotopic dating of zircon was reported by VINOGRADOV et al. (1952) and TILTON et al. (1955); their results demonstrated the feasibility of U-Th-Pb dating of accessory minerals from common rock types. The isotopic ages found in zircon are generally discordant with ^{206}Pb/^{238}U age $< ^{207}$Pb/^{235}U age $< ^{207}$Pb/^{206}Pb age. One such example of zircons with discordant ages is given in Fig. 2, which is reproduced from the study by SILVER and DEUTSCH (1963). Also see Table 4. The study was particularly useful in that the data lie along an episodic lead-loss line, and it was the first study that verified the episodic lead-loss hypothesis of WETHERILL. The data do not lie along continuous diffusion lines and, in fact, no clear example has yet been found to support the continuous diffusion hypothesis of TILTON (1960). A good example of episodic lead loss (and uranium gain) is given in the study of DAVIS et al. (1968) as presented here in Figs. 3 and 4. The zircons are from the Front Range of Colorado from the Precambrian Idaho Springs Formation adjacent to the Tertiary Eldora stock and from the Precambrian Silver Plume Granite adjacent to the Tertiary Jamestown stock. In this case, the episodic event was intrusion of the stocks. Hydrothermal experiments on a metamict Ceylon zircon in 2 molal NaCl at 500° C and 1,000 bars (PIDGEON et al., 1966; this report Fig. 5a, b) also showed a very rapid episodic lead loss of approximately 50 percent of the lead in 11 hours. Uranium content was little affected. Weathering also can apparently cause discordance in U-Pb ages of zircon (STERN et al., 1966) as shown in Fig. 6. Also see data in Table 4. The conclusion from these studies is that zircons are usually discordant and that this discordance may be caused rapidly by geologic events usually not considered to be strong. Also important is the fact that usually only part of the lead is lost. If the Wetherill concordia diagram is used, reliable mineral ages can be obtained. Although zircons easily lose some lead, they are remarkably resistant

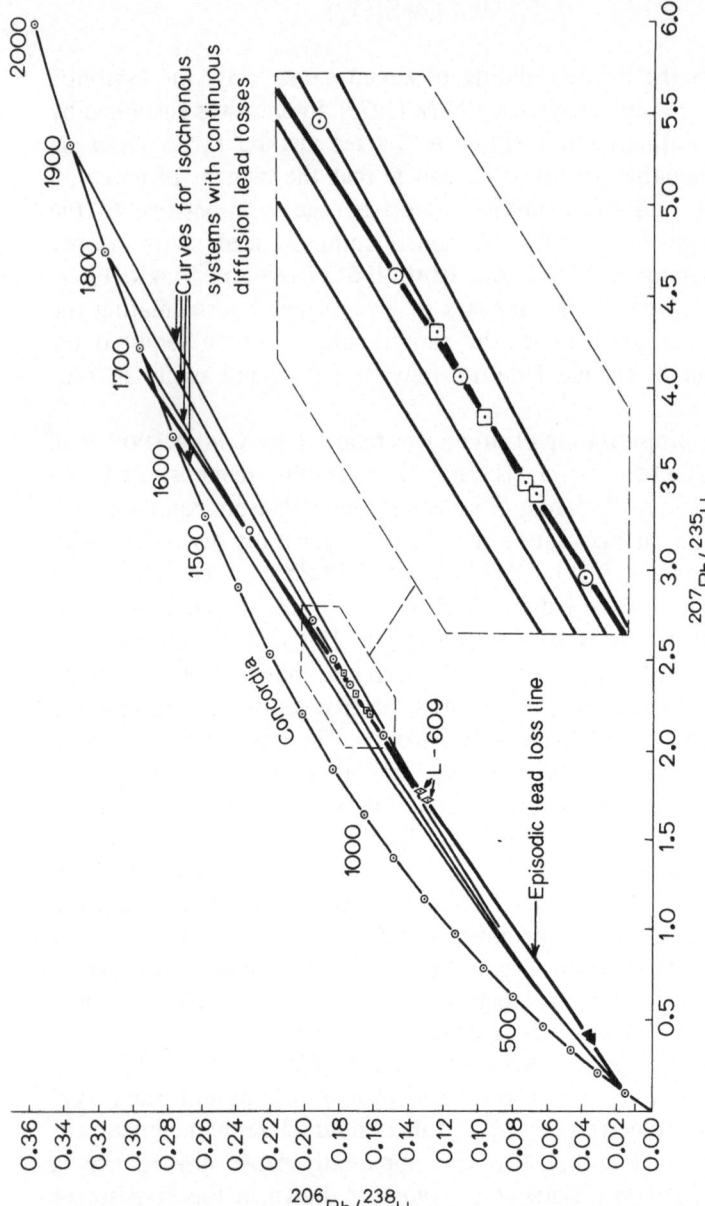

Fig. 2. Concordia plot for zircons from the Precambrian Johnny Lyon Granodiorite (SILVER and DEUTSCH, 1963). Triangles are two samples of uranothorite washes. Squares are four size fractions of zircons from one sample. Circles indicate various partial fusions and an average value of the intermediate zone of the retained 200-mesh fraction. The diamonds indicate a separate sample of zircon collected about 3 miles from the first sample. The inset is a photographic enlargement (5x) (from SILVER and DEUTSCH: Uranium Lead Isotopic Variations in Zircons. The Journal of Geology, Vol. 71, No. 6, pp. 721–758. Copyright by the University of Chicago 1963).

to total lead loss unless they are subjected to the processes of total recrystallization or complete solution and redeposition. At least three studies give isotopic verification of zircon xenocrysts in igneous rocks. STERN *et al.*(1965) reported a $^{206}Pb/^{238}U$ age of 494 m.y. for zircons in a Tertiary diorite porphyry from the La Sal Mountains, Utah, which

Table 4. *U-Th-Pb data for selected zircons*

Age of zircons (million years)	Geologic unit	Locality	Element concentration (ppm)			Atomic ratios			References
			U	Th	Pb	$^{206}Pb/^{204}Pb$	$^{207}Pb/^{204}Pb$	$^{208}Pb/^{204}Pb$	
3,500	Morton Gneiss, fresh granite	Minnesota	962	—	548	1961	465	147.5	Catanzaro (1963)
			767	—	478	2703	689	188.5	Catanzaro (1963)
			1.136	—	374	699	90.6	153.6	Catanzaro (1963)
	Morton Gneiss, residual clay	Minnesota	1.142	—	104	546	133.1	88.9	Stern et al. (1966)
			339	—	126	856	224.0	138.0	Stern et al. (1966)
			406	—	139	1055	300.1	136.1	Stern et al. (1966)
2,900	Fresh granite gneiss	Bighorn Mountains, Wyoming[a]	580	—	288	1068	219.7	141.1	Heimlich and Banks (1968)
			595	—	300	1041	214.5	144.4	Heimlich and Banks (1968)
			533	—	285	1237	253.4	187.4	Heimlich and Banks (1968)
1,930	Fresh granite gneiss	Little Belt Mountains, Montana	2227	1421	261	229	36.6	59.1	Catanzaro and Kulp (1964)
			549	217	138	694	90.1	126.9	Catanzaro and Kulp (1964)
			1417	276	216	658	90.1	95.2	Catanzaro and Kulp (1964)
			553	159	115	1042	127.2	121.3	Catanzaro and Kulp (1964)
			710	346	123	629	81.4	106.7	Catanzaro and Kulp (1964)
			1047	338	176	303	46.4	74.2	Catanzaro and Kulp (1964)
1,670	Johnny Lyon Granodiorite Size fractions (microns)	Arizona[a,b]							
	150—75		510	—	100	862	99.0	191.8	Silver and Deutsch (1963)
	75—52		542	—	103	894	101.1	205.1	Silver and Deutsch (1963)
	52—37		569	—	103	1113	123.1	250.2	Silver and Deutsch (1963)
	37—20		579	—	107	1139	125.2	263.4	Silver and Deutsch (1963
1,400	Leeuwfontein Syenite	South Africa	408	415	74.7	516	59.9	163.8	Oosthuyzen and Burger (1965)

[a] Lead contents are radiogenic values only. — [b] Sample L-132.

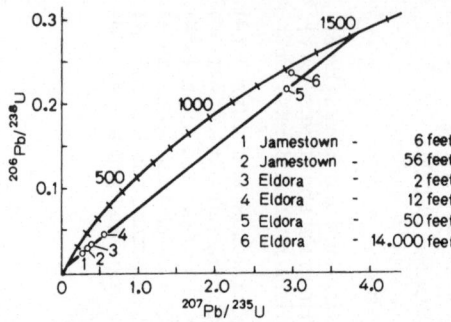

Fig. 3. Concordia diagram for zircons from the Precambrian Idaho Springs Formation adjacent to the Tertiary Eldora stock and from the Precambrian Silver Plume Granite adjacent to the Tertiary Jamestown stock, Colorado (DAVIS *et al.*, 1968)

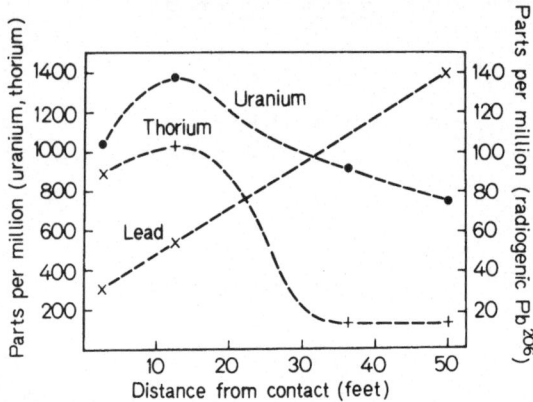

Fig. 4. Change in concentration of uranium, thorium, and lead in zircon with distance from the Eldora stock, Colorado (DAVIS *et al.*, 1968)

bears amphibolite xenoliths. PEARSON *et al.* (1962) reported a Pb-α age of 530 m.y. for zircons in the Tertiary Lincoln Porphyry of the Leadville district, Colorado. PASTEELS (1964) found a $^{207}Pb/^{206}Pb$ age of 385 m.y. for zircons in the xenolith-bearing Hercynian granite of Monti Orfano (age of about 280 m.y.) in the Alps. In all three areas, the granites are poor in zircon (Oral commun., 1967, for samples analyzed by PEARSON *et al.*; written commun., March 1968, for those analyzed by PASTEELS) and are small bodies in which crystallization times may be short. Zircon xenocrysts from larger, more deep-seated, more slowly crystallizing bodies may possibly be completely reset even in zircon-rich granites.

The mechanism, which causes the discordant ages in zircon, is not yet known. SILVER (1963) showed a relationship between uranium content and $^{206}Pb/^{238}U$ age in cogenetic zircons. This relation suggests

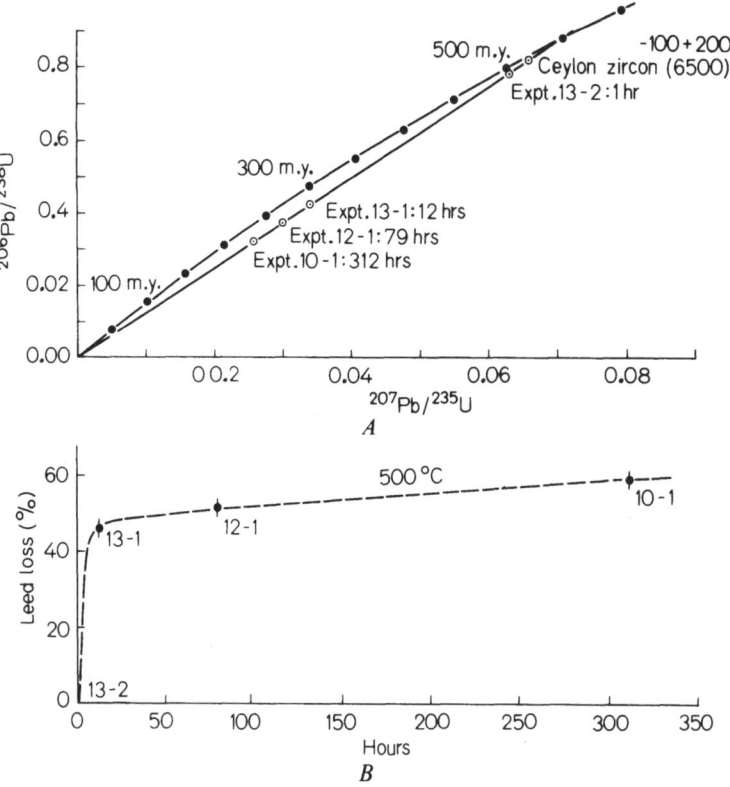

Fig. 5. Effects of loss of lead from Ceylon zircon (6,500) (÷100 to +200 mesh) as a result of hydrothermal experiments at 500° C in 2 molal NaCl at 1,000 bars fluid pressure: A, Concordia plot showing episodic lead loss: B, change of lead content in zircon with time (from PIDGEON, O'NEIL and SILVER: Uranium and Lead Isotopic Stability in a Metamict Zircon under Experimental Hydrothermal Conditions. Science **154**, 1538—1540. Copyright 1966 by the American Association for the Advancement of Science)

that a time-dependent diffusion constant (such as radiation damage) may be the answer; however, many exceptions to this relationship exist. A most promising approach seems to be consideration of the formation of domains rich in H₂O, heavy trace elements (including uranium and thorium), and yttrium, such as discussed by GRÜNEN-FELDER (1963), GRÜNENFELDER et al. (1964, 1968), and STEIGER and WASSERBURG (1966). Whether this formation of domains is a continuous process or is assisted by catastrophism has not yet been determined. Support for the importance of GRÜNENFELDER'S "multiphase" approach to the dating of zircons has been found by STEIGER and WASSERBURG (1969). They have found examples where the secondary age (lower intersection of the discordia line) determined when an episodic model

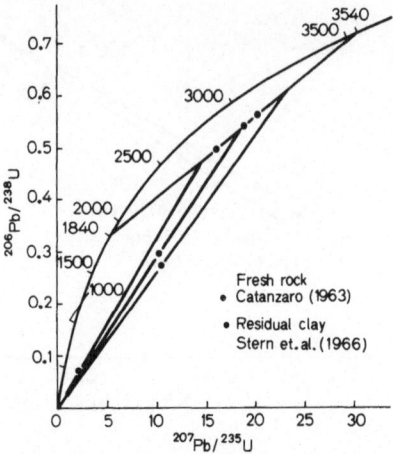

Fig. 6. Concordia diagram of zircon samples from fresh Morton Gneiss (CATANZARO, 1963) and its residual clay (STERN, GOLDICH, and NEWELL, 1966), Minnesota

is assumed disagrees with that expected from other information. Their argument is convincing that some zircons behave like mixtures of two phases – one that is nearly concordant and another that is highly discordant in a manner expected in models involving some sort of continuous diffusion.

A key step in the zircon purification procedure is use of a leach with hot HNO_3 to remove pyrite, apatite, uranothorite, and other impurities. SILVER and DEUTSCH (1963) suggested that the leach is neccessary to remove all the uranothorite. Several studies have been made to try to evaluate the effects, if any, of this leaching on the U-Th-Pb dating of the zircon. SILVER and DEUTSCH (1963) and STEIGER and WASSERBURG (1968) concluded that leaching either with the normal method using HNO_3 or with aqua regia does not seriously affect the U-Pb dating. Ground zircon boiled in aqua regia may preferentially lose [208]Pb, (STEIGER and WASSERBURG, 1968). The leachings are always radiogenic and some investigators have used them in dating by the [207]Pb/[206]Pb isochron method (OOSTHUYZEN and BURGER, 1965; CLIFFORD et al., 1962; PASTEELS, 1964) and have reported good agreement with ages determined by other methods. No concentrations of uranium, thorium, and lead are yet published that would help to evaluate these leaches on concordia diagrams. STEIGER and WASSERBURG (1969), however, were able to make full determinations on the U-Th-Pb system on strong acid leaches from zircons subsequent to hand-picking and two prior stages of weaker acid leaches! They conclude that they were actually dissolving a highly discordant phase of the zircon and, from inspection of the data on concordia diagrams that the discordance occured primarily through continuous diffusion.

Long-term disequilibrium of intermediate daughters could pose a problem in dating of zircons; however, MILLARD (1963) and PASTEELS (1965) found ^{210}Pb to be in radioactive equilibrium, and DOE and NEWELL (1965) found only a small disequilibrium in ^{234}U in a few zircons. Disequilibrium does not appear to be a major problem.

There is the temptation to determine just the lead isotopic composition of radioactive minerals and obtain an age measurement from the ^{207}Pb-^{206}Pb isochron diagram. The advantages of this method are that the concentration determinations are eliminated, but, of more scientific value, any episodic changes in $(^{238}U/^{204}Pb)_n$ at time close to zero $(t' \to o)$ do not affect the age calculation, whereas, in the concordia treatment, such alterations could destroy the usefulness of the method if there was also an alteration at $t \gg o$. The reverse situation holds whenever the product of the age of the alteration episode (t') and value of $(^{238}U/^{204}Pb)'_n$ established by the episode becomes significant. Whenever this condition holds true, the ^{207}Pb-^{206}Pb isochron becomes of less value and in fact straight lines will be maintained only under special conditions. Even if t' is known, $(^{238}U/^{204}Pb)'_n$ could have changed so irratically such as to destroy any linear relationship. The episodic alteration ages usually obtained from concordia are near enough to $t' = o$ (~ 100 m.y.) so that the error introduced in dating Precambrian zircons by the ^{207}Pb-^{206}Pb isochron method is not large (20–40 m.y.), but may reach 10 percent of the concordia age if t' is much greater. However, the errors become more significant in younger rocks and may reach 30 percent in rocks of Paleozoic age. The resultant isochron age is usually too young because normal discordance is the usual case. The concordia treatment, however, is valuable at any time there is an episodic alteration condition, such that the product $(^{238}U/^{204}Pb)'_n \cdot (e^{\lambda_8 t'} - 1)$ is significant. The age of the alteration may in fact be determined.

Th-Pb ages on zircons usually, but not always, range from the ^{206}Pb/^{238}U age to the ^{207}Pb/^{206}Pb age. The Th-Pb ages are rigorously interpretable in some cases by use of the U-Th-Pb concordia diagram (STEIGER and WASSERBURG, 1966) but are generally unreliable for dating work.

Evaluation. Zircons are usually discordant in their age patterns. Reliable ages by U-Pb methods may be obtained, however, through analyses of several fractions of zircon from one rock sample or of zircons from several rock samples by treating the data with the concordia-discordia method of WETHERILL. Some problem may be encountered in this method when trying to date zircon-poor intrusions suspected of containing xenocrysts. The zircons in these intrusions, however, may prove useful as a tracer of zircon xenocrysts. Th-Pb dating of zircons is not reliable.

Though some lead is easily lost from zircons as a result of geologic events such as metamorphism, complete resetting rarely occurs and the concordia-discordia treatment furnishes a powerful "eye" to see through metamorphic events to determine rock ages.

2. Sphene (CaTiSiO₅)

As sphene is one of the most widespread radioactive accessory minerals in igneous and metamorphic rocks (it is particularly abundant in amphibolite), the lack of more extensive dating of this mineral is surprising. Sphene was first demonstrated to be datable by VINOGRADOV et al. (1952). (Those wishing to see early Russian data on accessory minerals in English translation are referred to VINOGRADOV, 1956.) TILTON et al. (1955) also early demonstrated the feasibility of dating sphenes where it is in "normal" abundance by the U-Th-Pb method. The radiogenic enrichment of most sphenes is not as large as that of zircon so that common lead corrections are more important. As few data are yet available, the whole list of reports may be easily studied. Other key papers on sphene are by ZHIROV et al. (1957), BURGER et al. (1965), OOSTHUYZEN and BURGER (1965), and TILTON and GRÜNENFELDER (1968); the last-mentioned contains 13 sphene analyses which are compared with analyses of zircons. The comparison of sphene and zircon is given in Fig. 7, which is based on the data in Table 5. The sphene appears to be more concordant in its age values than does the

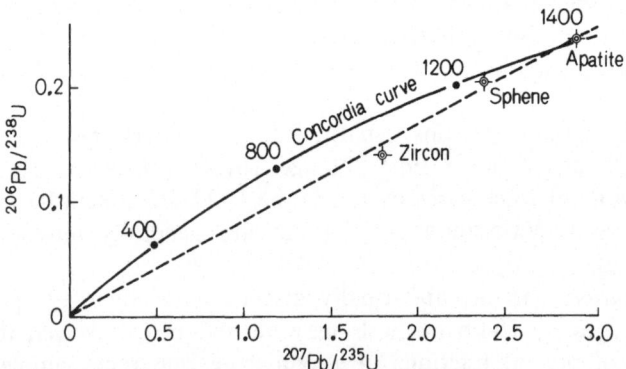

Fig. 7. Concordia diagram for Leeuwfontein Syenite, South Africa, showing isotopic relations of zircon, sphene, and apatite (from OOSTHUYZEN and BURGER, 1965)

Table 5. Data for U-Th-Pb dating of sphenes and comparison of ages on some sphenes with those on some coexisting zircons

Location	Phase	Element concentrations (ppm)			^{206}Pb/^{204}Pb	^{207}Pb/^{204}Pb	^{208}Pb/^{204}Pb	Comment	References
		U	Th	Pb					
A. U-Th-Pb data on some selected sphenes									
Leeuwfontein Syenite, South Africa	Sphene	63.8	230	32.1	121.4	24.47	149.5	Sphene is more concordant than zircon	OOSTHUYZEN and BURGER (1965)
	Leach of sphene concentrate	—	—		28.49	16.70	50.48	Spene is more concordant than zircon	OOSTHUYZEN and BURGER (1965)
Granite, Sandia Mountains, New Mexico, USA	Sphene, dark / light	148.9 148.8	— —	31.93 31.67	259.1 350.9	36.74 46.18	282.9 315.8	Sphene concordant at 1450 m.y. (zircon is discordant), which is probable rock age	TILTON and GRÜNENFELDER (1968)
Granite gneiss along French R. Ontario, Canada	Sphene	145.1 90.2	— —	20.46 12.83	161.8 277.8	26.44 34.83	97.2 118.5	Sphene nearly concordant at metamorphic age of 1,000 m.y.; zircon reflects older age	TILTON and GRÜNENFELDER (1968)

2*

Table 5 (continued)

Location	Phase	Apparent ages (X·10^6 yr)				References
		^{206}Pb/^{238}U	^{207}Pb/^{235}U	^{207}Pb/^{206}Pb	^{87}Sr/^{87}Rb	
B. Comparison of ages on sphene with those on zircon						
Granite, Sandia Mountains, New Mexico, USA	Sphene	1450	1465	1480	—	TILTON and GRÜNENFELDER (1968)
	Sphene	1440	1460	1490	—	TILTON and GRÜNENFELDER (1968)
	Zircon	935	1105	1460	—	STEIGER and WASSERBURG (1966)
	Zircon	1290	1360	1470	—	STEIGER and WASSERBURG (1966)
	Whole rock	—	—	—	1400	WASSERBURG et al. (1965)
Granite gneiss along French R., Ontario, Canada	Sphene, dark	989	1024	1100	—	TILTON and GRÜNENFELDER (1968)
	Sphene, light	998	1018	1060	—	TILTON and GRÜNENFELDER (1968)
	Zircon	1220	1315	1470	—	T.E. KROGH (personal commun., in TILTON and GRÜNENFELDER, 1968)
	Whole rock	—	—	—	1725	HART et al. (1967)

companion zircon. OOSTHUYZEN and BURGER (1965) determined the Th-Pb age of their sphene as 1,190 m.y., whereas that age on the zircon is 705 m.y. The Th-Pb age on sphene may be found to be more reliable than that on zircon.

The major evaluation study of TILTON and GRÜNENFELDER (1968) (analytical data for selected suites are given in Table 5) indicates that sphenes will usually give concordant ages but at either end of a discordia line. They probably will not be as good an "eye" to see through metamosphism in obtaining rock ages. This characteristic is illustrated by the analyses of TILTON and GRÜNENFELDER on granite-gneiss from along the French River in Ontario. The whole-rock Rb-Sr isochron age of this gneiss is given by HART et al. (1967) as 1,725 m.y., whereas Rb-Sr mineral ages such as those for biotite reflect the age of the so-called Grenville orogeny at about 1,000 m.y. Two sphene fractions were a little discordant but gave $^{207}Pb/^{206}Pb$ ages of 1,060 and 1,100 m.y. In contrast, the zircon (J. E. KROGH, personal comm., in TILTON and GRÜNENFELDER, 1968) again demonstrated the difficulty of completely resetting the zircon in nature as it has a $^{207}Pb/^{206}Pb$ age of 1,470 m.y. Unfortunately, TILTON and GRÜNEN-FELDER did not report the thorium contents of their sphenes so further evaluation of Th-Pb dating is not possible at this time.

Little laboratory experimental work has yet been done on sphene. OOSTHUYZEN and BURGER (1965) did analyze the isotopic composition of the acid leach solution of their sphene (Table 5) and found it to be somewhat radiogenic; however, no more so than is found in studies of acid leachings of zircon. (See section on zircon.) BURGER et al. (1965) also reported values on leachings of sphene from 600-m.y.-old rocks which may be as radiogenic as the sphene ($^{206}Pb/^{204}Pb$ of 138.2 of leach Ll from sphene X; the sphene has a value of 114.3). The effect of this leaching on the sphene is not yet completely evaluated.

A need exists for additional laboratory work, such as hydrothermal experiments, disequilibrium studies, and petrographic studies which might explain how an investigator can tell independently from other dating work just which kind of age (metamorphic or rock age) might be expected from sphene. Existing data suggest that radon leakage is not a major problem.

3. Phosphates

Apatite $(Ca_5FP_3O_{12})$. Apatite is another mineral, which has been little used and investigated for dating but which has been shown to be of possible interest. The first apatites were dated by ALDRICH et al. (1955) and TILTON et al. (1955). Other studies that

Table 6. *U-Th-Pb data for apatites*

Rock Type	Locality	Element concentrations (ppm)			Atomic ratios				References
		U	Th	Pb	²⁰⁴Pb	²⁰⁶Pb/²⁰⁴Pb	²⁰⁷Pb/²⁰⁴Pb	²⁰⁸Pb/²⁰⁴Pb	
						Atomic abundances			
						²⁰⁶Pb	²⁰⁷Pb	²⁰⁸Pb	
Leeuwfontein suite	South Africa	20.7	—	28		38.16	17.60	83.58	OOSTHUYZEN and BURGER (1965)
Granite	Tory Hill, Ontario, Canada	90.5	—	136		31.9	16.3	76.3	TILTON et al. (1955)
Granite	Umcompaghre uplift, Colorado	12.43	—	3.14[a]		41.07	17.78	52.16	ALDRICH et al. (1955)
Authigenic apatite from upper formation of Krivoirog Series	Ukraine, USSR	1020	—	320	0.08	86.27	10.81	21.84	TUGARINOV et al. (1963a)
		1000	—	370	0.03	64.02	6.95	26.91	TUGARINOV et al. (1963)
Metasomatic rocks, Sutam Series	Southern Aldan Shield, Siberia	13200	—	3110	0.07	88.60	8.92	2.41	ISKANDEROVA and LEGIERSKY (1966)
		13100	—	2960	0.04	89.73	8.84	1.39	ISKANDEROVA and LEGIERSKY (1966)
		1360	—	360	0.21	82.15	10.36	7.28	ISKANDEROVA and LEGIERSKY (1966)

[a] ppm of ²⁰⁶Pb only.

have included apatite dating are by VOLOVYEV *et al.* (1963), TUGARINOV *et al.* (1963b), OOSTHUYZEN and BURGER (1965), and ISKANDEROVA and LEGERSKIY (1966).

A comparison of apatite (Table 6) with zircon (Table 4) and sphene (Table 5) is given in Fig. 7 (from OOSTHUYZEN and BURGER, 1965). The apatite is nearly concordant but has slight reverse discordance which, because the radiogenic enrichment is not large, might be due to the common lead correction. ISKANDEROVA and LEGIERSKIY (1966) reported three unusually radiogenic apatites from the Sutam series in the Aldan shield as concordant near 1,400 m.y.; they report this age to be in good agreement with the known geologic age.

Xenotime (YPO_4). Another phosphate of some dating interest is xenotime (IVANTISHIN *et al.*, 1962; KOMLEV *et al.*, 1961; LYAKHOVICH, 1961).

Monazite ($CePO_4$) has received enough attention to warrant discussion in a section by itself (see 6.).

4. Pyrochlore ($NaCa(Cb, Ta)_2O_6F$)

Pyrochlore, or microlite, which is also an ore of niobium, has received some dating interest (ALDRICH *et al.*, 1956; LYAKHOVICH, 1961; MIRKINA and ISKANDEROVA, 1962; MIRKINA *et al.*, 1962; ZYKOV *et al.*, 1964; MIRKINA and MAKAROCHKIN, 1966). The mineral may show reverse discordance and has yet to be completely evaluated. One example is shown (Fig. 9 in the monazite section) by ZYKOV *et al.* (1964). As the pyrochlore lies well along the discordia line which connects the mean value of the zircons with a highly reverse discordant monazite, the reverse discordance of the pyrochlore and monazite appears to be caused episodically; if the reverse discordance were caused by continuous diffusion of uranium from (or radiogenic lead into) the pyrochlore, the line connecting pyrochlore and monazite would lie below the probable discordia line of maximum slope through the zircons (i.e., a line extending through zero). A better developed zircon discordia line is needed to help verify the nature of the reverse discordance.

5. Epidote Group ($Ca_2Al_2OHSi_3AlO_{12}$)

Epidote proper has received little attention for U-Th-Pb dating, but some data which suggest that it should be investigated further have been published (NARBUTT *et al.*, 1959; MINEEV, 1959). Considerable work, however, has been done on allanite (the cerium-bearing member). The first analysis was reported by VINOGRADOV *et al.* (1952), and the

Table 7. *U-Th-Pb data on some selected allanites from pegmatites*

Location	Element concentrations (percent)			Atomic ratios							Reference
	U	Th	Pb	^{206}Pb/^{204}Pb	^{207}Pb/^{204}Pb	^{208}Pb/^{204}Pb					
							Atomic abundance				
							^{204}Pb	^{206}Pb	^{207}Pb	^{208}Pb	
Kola Peninsula, U.S.S.R.	0.008	0.66	0.060	30.43	16.41	666.7					Zykov et al. (1964)[a]
Northern Sweden	3.65	1.40	1.04				0.09	79.34	9.23	11.34	Welin and Blomqvist (1964)
Northern Sweden	0.072	1.74	0.11				0.03	15.56	2.10	82.31	Welin and Blomqvist (1964)

[a] Concentrations by radiochemical method and volumetrically by the hydrosulfite-phosphate-vanadate method for uranium; radiochemically and colorimetrically with arsenozo III for thorium; and by oscillographic polarigraph for lead.

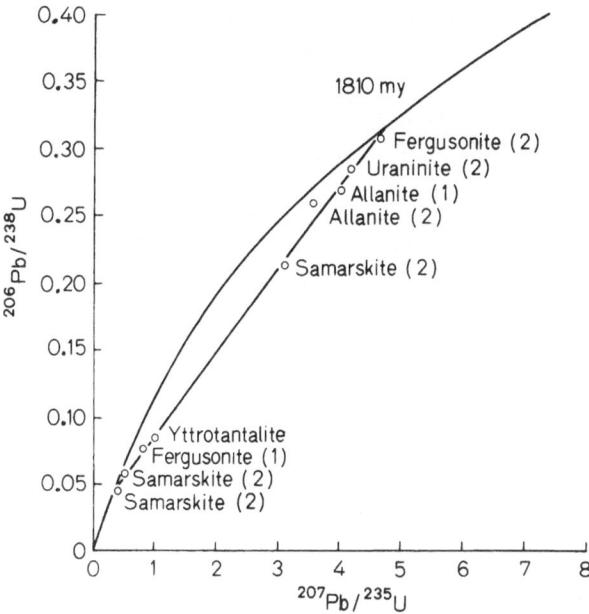

Fig. 8. Concordia diagram of the data for minerals from pegmatites from Sweden (after WELIN and BLOMQVIST, 1964). Minerals labeled (1) are from northern Sweden; those labeled (2) are from central Sweden

allanite was found to have reverse discordance. (See VINOGRADOV, 1956, for data in English translation.) VINOGRADOV (1956) reviews a total of eight analyses, two of which displayed reverse discordance. Many allanites and orthites are not strongly radiogenic and good comparisons of data on allanite and orthite with those on zircons are not available. Some evaluation of U-Th-Pb dating on this mineral family was presented by WELIN and BLOMQVIST (1964) (see Table 7, Fig. 8) and MARKINA and MAKAROCHKIN (1966). The two allanites from Sweden (WELIN and BLOMQVIST, 1964) have very different contents of uranium and values of Th/U, yet both allanites are of about the same discordance. The allanite richer in uranium is slightly the more discordant.

Additional laboratory and petrographic studies are needed to help determine what causes some allanites to have reverse discordance and to find criteria to help predict this situation. A paper on differential lead extraction from allanite exists (IORDANOV, 1960).

6. Monazite (CePO$_4$)

Monazite was the first non-ore mineral after the cyrtolitic variety of zircon to be dated by the U-Th-Pb method (NIER et al., 1941). VINOGRADOV's review paper (1956) lists 19 analyzed monazites from

Russia alone. Next to uranium minerals and zircon, monazite receives the greatest emphasis today. Although monazite usually has normal discordance of $^{206}Pb/^{238}U$ age $< ^{207}Pb/^{235}U$ age $< ^{207}Pb/^{206}Pb$ age (strongly so in some specimens), some examples are known to have strong reverse discordance. (For one such example see Fig. 9 and Table 8.) Why this reverse discordance occurs in occasional monazites is not understood, but the theoretical base provided by WETHERILL (1963) allows distinction of continous diffusion of uranium from episodic loss of

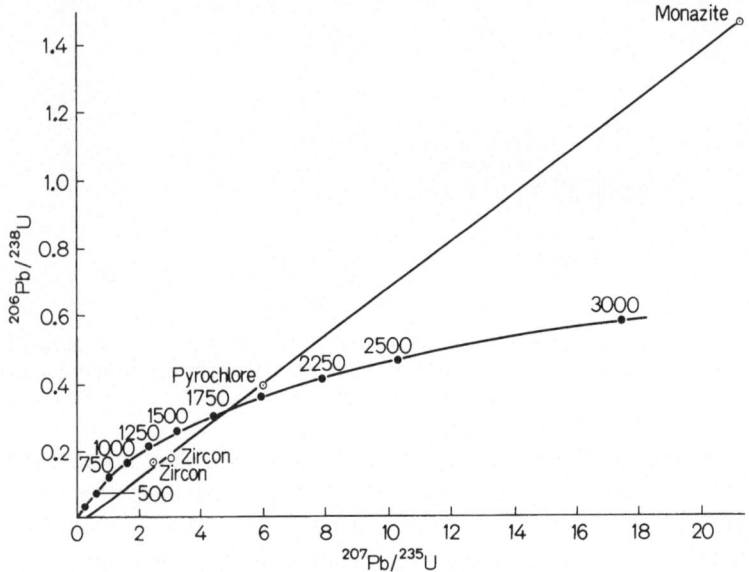

Fig. 9. Concordia diagram of pegmatites and alkalic intrusives of Marealbian time from the Kola Peninsula (after ZYKOV et al., 1964). Ploting of data for allanite from Table 7 is omitted as it is exceedingly sensitive to the common lead correction because of small radiogenic enrichment

uranium. (See 4., *Pyrochlore* section.) Much emphasis has been placed on acid-leaching experiments (TILTON and NICOLAYSEN, 1957; STARIK et al., 1958d; STARIK et al., 1960; BURKSER et al., 1962; KOMLEV et al., 1964; STARIK and LAZAREV, 1964; BURGER et al., 1965, 1967), which have shown that some uranium and uranogenic lead are leached relatively easily in the laboratory leaching and that with Th and ^{208}Pb are less easily leached. These studies suggest the likelihood of episodic losses in nature also. Examples of monazites with reverse discordance, which have well-established zircon isochrons, are needed to better evaluate this conclusion.

Because the Th-Pb system is less easily leached in laboratory studies, STARIK et al. (1960) concluded that the Th-Pb age is probably

the most reliable of all ages. A recent investigation by BURGER *et al.*
(1967) indicates that leaching in nature does not mirror the results of the
laboratory experiments on normally discordant monazites, and the
concordia treatments of the U-Pb system on these monazites gives the
best age estimate. They also point out that it is apparent from modern
detailed studies where independent evidence of age exists that the
Th-Pb age is usually too low.

Examples are also known where monazite is nearly concordant
(Fig. 10, Table 8). In the Rocky Mountain region monazite is a
relatively abundant mineral and monazites appear to have a tendency
to be concordant which would provide a reasonable dating tool.

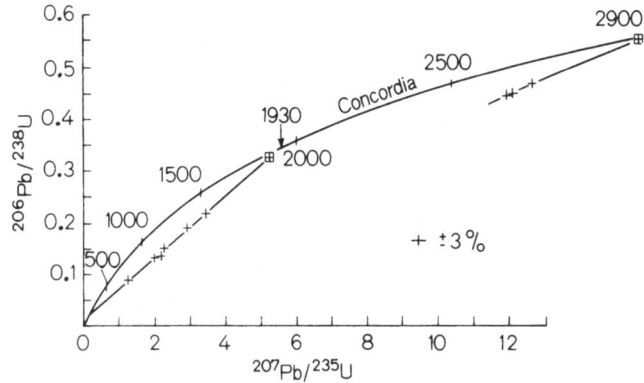

Fig. 10. Concordia diagram showing monazite and zircon relations. Discordia
line intersecting concordia at 1,930 m.y. is from CATANZARO and KULP (1964),
Little Belt Mountains, Montana; The discordia line intersecting concordia at
2,900 m.y. is from HEIMLICH and BANKS (1968), Bighorn Mountains, Wyoming.
Crosses are plottings for zircons; boxes are plottings for monazites

Little rigorous evaluation has been made on the effect of metamorphism on monazite so an estimate of the ability of this mineral to
retain some isotopic identity through periods of metamorphism is not
known. CATANZARO and KULP (1964) speculate that, on the basis of
geographic relations and lithologic similarity with older rocks, both
the zircon and the monazite in the Little Belt Mountains are indeed
completely reset by metamorphism at 1,930 m.y. There is no independent study that suggests that this complete resetting is possible
in zircons, short of melting. Monazite remains to be studied from this
aspect.

Some special types of intermediate continuous daughter loss could
cause either normal or reverse discordance but no studies were found
dealing with this. STARIK *et al.* (1960) do report, however, that radium
and radon in the uranium decay chains may be leached from the crystal

Table 8. Selected U-Th-Pb data for monazite and pyrochlore

Mineral and rock type	Locality	Element concentrations (percent)			Atomic ratios			Reference
		U	Th	Pb	$^{206}Pb/^{204}Pb$	$^{207}Pb/^{204}Pb$	$^{208}Pb/^{204}Pb$	
Monazite, pegmatoidal granite	Kola Peninsula, U.S.S.R.	0.65	19.6	3.35	364.1	52.3	1011.1	ZYKOV et al. (1964)[a]
Pyrochlore, alkalic pegmatites	Kola Peninsula, U.S.S.R.	0.935	0.265	0.385	1388.7	170.2	106.8	ZYKOV et al. (1964)
Monazite, granite gneiss	Little Belt Mountains, U.S.A.	0.213	8.84	0.802	312.5	50.2	35.4	CATANZARO and KULP (1964)[b]
Monazite, granite gneiss	Bighorn Mountains, U.S.A.	0.0830	—	0.998[c]	206.3	53.8	4521	HEIMLICH and BANKS (1968)[b]

[a] Concentrations by radiochemical method and volumetrically by the hydrosulfite-phosphalte-vanadate method for uranium; radiochemically for thorium; and by oscillographic polarigraph for lead.
[b] Concentrations by isotope dilution.
[c] Radiogenic lead only.

and that ^{224}Ra in the thorium decay chain is leached less. In view of this, further studies of loss of intermediate daughters seem advisable.

Evaluation. Monazite is a common accessory and studies done so far suggest that it can be of great value in dating. In some regions, such as the Rocky Mountains, monazites tend to be concordant; in other they are of normal discordance and are adequately treated by concordia diagrams. A few monazites have reverse discordance, the cause of which can not be well explained from available data. Some controlled studies are needed to determine how well the U-Pb system in monazite survives metamorphism. Because some uranium and uranogenic lead are easily removed from the monazite in laboratory acid-leaching studies, some effect from metamorphism is expected and the mineral may be completely reset in some cases. The Th-Pb ages appear to be unreliable.

7. Uranium Minerals

Ore minerals of uranium were of course among the first minerals for which lead isotopic compositions were determined (ASTON, 1929). Though ore minerals of uranium and thorium tend to be sparse in common rock types, about seven papers per year that date them have continued to be published since 1960. Uraninite from sulfide-free environments such as exist in many pegmatites is most concordant in its uranogenic lead ages and has played a key role in geochronology. Examples are presented in Table 9. The most widely used value of the ^{87}Rb decay constant was determined by calibration of Rb-enriched minerals against coexisting uraninite from pegmatite (ALDRICH et al., 1956) as well as used in verifying the decay constants of ^{40}K (WETHERILL et al., 1956). WASSERBURG and HAYDEN (1955) have pointed out that the Th-Pb ages appear to be consistently lower by about 10 percent than the U/Pb ages.

As PbO_2 (a form that would match UO_2) is not a stable form of lead in nature, one might wonder why lead is not lost from uraninites even under sulfide-free environments. The lead might be in the form of PbO, which would account for the excess oxygen in most uraninites, and BERMAN (1957) proposed exsolution of oriented monomolecular layers of orthorhombic PbO along the cube planes in the cubic uraninite, because the a_0 and c_0 of PbO is similar to that of UO_2. When sulfur is present, PbO is no longer stable and the lead exsolves as galena, often beyond the grain boundaries. KOEPPEL (1968) found that hematite and selenides have effects on pitchblende similar to the effect of sulfur but that often whole-rock analyses of milligram quantities of pitchblende (similar to uraninite but with no thorium) will often give con-

Table 9. *Dating of uraninites with concordant ages by U-Th-Pb method*

Locality	Element abundance (weight percent)			Atomic abundance				Age (m.y.)			Ref.
	U	Th	Pb	^{204}Pb	^{206}Pb	^{207}Pb	^{208}Pb	$^{206}Pb/^{238}U$	$^{207}Pb/^{235}U$	$^{208}Pb/^{232}Th$	
Viking Lake, Beaverlodge dist., Canada	52.9	5.24	17.14	0.005	87.370	10.091	2.534	1850	1880	1670	1
Blackstone Lake pit, Conger Twp., Parry Sound, Canada	69.3	2.99	10.72	0.007	91.956	6.658	1.379	994	993	897	1
Pit in Con. XVI Cardiff Twp., Canada	62.3	6.61	9.92	0.013	90.119	6.750	3.118	1000	1020	870	1
Wilberforce, Canada	53.52	10.37	9.26	0.009	87.983	6.589	5.419	1077	1035[a]	983	2
Parry Sound, Canada	69.0	2.95	10.8	0.006	91.805	6.794	1.395	1003	1030[a]	945	3
Strickland quarry, Portland, Conn., USA	79.1	2.98	3.07	0.013	93.542	4.948	1.497	268	266	239	1

References: (1) WASSERBURG and HAYDEN (1955); (2) NIER (1939); (3) NIER *et al.* (1941).

[a] $^{207}Pb/^{206}Pb$ age.

cordant ages because the lead has not migrated a great distance. KOEPPEL (p. 56) also felt that many discordant pitchblendes are caused by the presence of the mineralization in cracks and fissures, which are affected by any tectonic rejuvenation. He also presented (p. 33) a possible example of continuous diffusion of lead from uraninite and thorite. Examples of discordant uraninites with exsolved lead as galena may be found in the studies of the Dominion Reefs by LOUW (1954) and HORNE and DAVIDSON (1955). The Dominion Reef uraninite data were the basis of WETHERILL's (1956a) graphical analysis of discordant U-Pb data and of many other analyses and studies (HOLMES and CAHEN, 1957; TUGARINOV, 1957, and prior papers; BURGER et al., 1962; and NICOLAYSEN et al., 1962).

STARIK et al. (1960) and BURKSER et al. (1962) determined the relative leachability of Th and Pb from pitchblende. STARIK et al. (1961) found ^{234}Th, the first decay product of ^{238}U, to be easily leached from uraninite. Similarly ^{228}Th is more easily leached than the parent ^{232}Th. ^{208}Pb is leached more easily than the uranogenic lead.

BURGER et al. (1962) considered the South African uraninites along with exsolved galena and concluded that the uraninites were completely reset about 2,000 m.y. ago without change in the morphology of the grains. Therefore, uraninite does not appear to be a good mineral with which to "see" through metamorphisms.

Evaluation. Under conditions of a sulfide-free environment and no strong subsequent metamorphism, pegmatitic uraninite and pitchblende are probably the most reliable dating minerals and give concordant ages. Under conditions of a relatively mild episodic event or in a sulfide-bearing environment, uraninite can often yield reliable concordia ages. Some evidence suggests that uraninite may sometimes be completely reset without change in grain morphology. The rarity of this mineral has restricted its use in dating to some special but very important experiments.

8. Other Minerals

Several other minerals, particularly those from pegmatites, have been used for dating. Other members of the columbite-tantalite family besides pyrochlore that have been dated include columbite-tantalite (ALDRICH et al., 1956b), tantalite (ZHIROVA et al., 1961), euxenite (ROBINSON et al., 1963), and, as plotted on Fig. 8, samarskite, yttro-tantalite, and fergusonite (WELIN and BLOMQVIST, 1964). LYAKHOVICH (1961) also reported data on fergusonite. Niobites and rutile (TiO_2) (CHOUBERT, 1964) and glauconite (mica structure) (POLEVAYA and PANTELEEV, 1962) have also received some attention.

Experiments on heating thorianite and uranothorite in vacuum have shown that the lead is more easily extracted from metamict material than from the crystalline mineral (ROBINSON *et al.*, 1963). See Fig. 2 for examples of uranothorite on a discordia plot from the study by SILVER and DEUTSCH (1963).

9. Whole Rock

Whole-rock dating is analytically difficult because of the lack of radiogenic enrichment in the lead, particularly ^{207}Pb. Though some approaches to whole-rock dating appear to have been made previously, the first concerted attempts were made by SOBOTOVICH (1961) on granites and gneisses about 3,000 m.y. old (for which ^{207}Pb effects would be relatively large) and by COBB (1961) on lower Paleozoic black shales and Swedish Kolm.

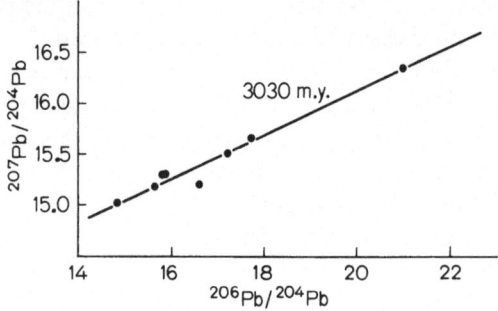

Fig. 11. Whole-rock plot of ^{207}Pb/^{204}Pb versus ^{206}Pb/^{204}Pb for rocks of granitic composition from the Taromskoe Quarry, Dnieper Region, Ukraine (SOBOTOVICH *et al.*, 1963a)

In a series of papers, SOBOTOVICH and his coworkers report success in dating very old rocks by the ^{207}Pb-^{206}Pb isochron method (Fig. 11, Table 10). In the study of the 3,000-m.y.-old granites and gneisses of the Taromskoe Quarry in the Ukraine (SOBOTOVICH *et al.*, 1963a), the Pb-Pb isochron was somewhat older than K-Ar ages on micas which are known to be easily affected by metamorphism. The determined values of uranium relative to lead were found to be too great for the observed lead isotopic compositions so that the ^{238}U/^{204}Pb-^{206}Pb/^{204}Pb age is much younger than the Pb-Pb age. Data on a rock sample as presented by ZARTMAN (1965) showed an opposite effect of much too little uranium. Zartman's other rocks appeared to be in fairly good U-Pb balance. Apparently the disturbances in U-Pb are recent. Many studies (SOBOTOVICH *et al.*, 1963b, c; and SOBOTOVICH and GRASHCHENKO, 1965) have dealt solely with the Pb-Pb age.

Table 10. *The isotopic composition of lead and uranium and lead contents from whole rocks, Taromskoe Quarry, Dnieper region, Ukraine, USSR* (from SOBOTOVICH et al., 1963a)

Speci-men No.	Rock type	Concentration (ppm)		Isotope ratios		
		U	Pb	$^{206}Pb/^{204}Pb$	$^{207}Pb/^{204}Pb$	$^{208}Pb/^{204}Pb$
95	Gray granite	—	—	17.20	15.50	37.70
102	Gray granite	—	—	16.60	15.20	36.60
103	Gray granite	1.85	9.0	15.66	15.18	36.22
117	Gray granite	2.30	14.5	15.80	15.30	37.00
105	Pink granite	—	—	14.82	15.01	34.00
106	Pink granite	—	—	17.70	15.65	38.30
107	Gneiss	4.3	13.0	21.00	16.35	41.50
112	Xenolith	1.9	12.0	15.85	15.30	35.40

The shales and the kolm analyzed by COBB (1961) plot in a manner that suggests loss of intermediate daughters. Measurements of ^{230}Th, ^{226}Ra, and ^{210}Pb indicate that ^{230}Th is nearly in equilibrium, ^{226}Ra is out of equilibrium, and ^{210}Pb is even more out of equilibrium. The disequilibrium is attributed to ground-water leaching. Phosphorite is another uranium-rich rock that has been investigated (GOLUBCHINA, 1962). TUGARINOV et al. (1963b) have made other investigations of U-Th-Pb dating of authigenic portions of some sedimentary rocks and have found some success in dating apatite-rich rocks.

GERLING and ISKANDEROVA (1966), in a review of some existing data on limestones, reported good agreement with existing ages in dating these rocks by $^{206}Pb/^{238}U$ isochrons. ISKANDEROVA (1966) also reported good results. If precise ages can be directly determined on limestones, exceedingly valuable correlations can then be made with paleontological specimens. Other work on 1,000-m.y.-old marbles from the Grenville Series near Balmat, New York (DOE, 1962a) was unsuccessful. The $^{206}Pb/^{238}U$ isochron line drawn through the plotting of the two whole-rock samples has a negative slope. Some marbles have values of $^{238}U/^{204}Pb$ that approach zero and WAMPLER and KULP (1962) have attempted dating these by common lead methods.

Evaluation. The method of determining ^{207}Pb-^{206}Pb age on whole-rock crystalline rocks shows great promise as a dating technique. The procedure has some capacity to "see" through metamorphisms and has

3 Doe, Lead Isotopes

the added feature of not being disturbed by recent weathering. Available studies in dating whole-rocks by other isochron methods demonstrate problems arising from recent changes in value of U/Pb.

Although some data suggest optimism in the use of $^{206}Pb/^{235}U$ isochrons on shales, limestones, and authigenic portions of clastic sediments, other information presents severe problems. The technique is not yet perfected.

III. Common Lead

1. General

"Common lead" is any lead from a phase with a low value of U/Pb and/or Th/Pb such that no significant radiogenic lead has been generated *in situ* since the phase formed. Such phases are galena and other sulfides such as pyrite, feldspars and in particular K-feldspar, micas, and most abundant rock types of Cenozoic age. Data on common lead are used in determining ages and, more important, in the solution of genetic problems. Presented herein, therefore, are data on those phases relevant to common lead applications even if sufficient radiogenic lead has been generated to markedly alter the initial isotopic composition of the lead such as in Precambrian crystalline rocks.

Relevant equations and parameters are in Tables 11 and 12.

Table 11. *Equations used in calculations concerning common lead*

1. Basic equation: $\dfrac{-dN}{dt} = \lambda N$

N = number of radioactive atoms
t = time
λ = constant of proportionality (decay constant)

2. Primary growth equations (all in atomic mass units)[a]:

Ratio at geologic time (t)	Ratio when earth formed[b]	Ratio in earth today	Decay from time earth was formed (T) to the geologic time of interest (t)
a) $\left(\dfrac{^{206}Pb}{^{204}Pb}\right)_t$ =	$\left(\dfrac{^{206}Pb}{^{204}Pb}\right)_T$ +	$\left(\dfrac{^{238}U}{^{204}Pb}\right)_n$ ·	$(e^{\lambda_8 T} - e^{\lambda_8 t})$.
b) $\left(\dfrac{^{207}Pb}{^{204}Pb}\right)_t$ =	$\left(\dfrac{^{207}Pb}{^{204}Pb}\right)_T$ +	$\left(\dfrac{^{235}U}{^{204}Pb}\right)_n$ ·	$(e^{\lambda_5 T} - e^{\lambda_5 t})$.
c) $\left(\dfrac{^{208}Pb}{^{204}Pb}\right)_t$ =	$\left(\dfrac{^{208}Pb}{^{208}Pb}\right)_T$ +	$\left(\dfrac{^{232}Th}{^{204}Pb}\right)_n$ ·	$(e^{\lambda_2 T} - e^{\lambda_2 t})$.

[a] For a close approach to a physical example see strataform ore deposits (Table 16 and Fig. 13) or the different approach given by KANASEWICH and FARQUHAR (1965) and KANASEWICH (1968a).

[b] Usually assumed to be the values in the troilite phase of meteorites (see MURTHY and PATTERSON, 1962; OVERSBY, 1970).

3*

Table 11 (continued)

3. Primary isochron equation[c]:

$$\frac{\left(\dfrac{^{207}Pb}{^{204}Pb}\right)_t - \left(\dfrac{^{207}Pb}{^{204}Pb}\right)_T}{\left(\dfrac{^{206}Pb}{^{204}Pb}\right)_t - \left(\dfrac{^{206}Pb}{^{204}Pb}\right)_T} = \left(\frac{^{235}U}{^{238}U}\right)_n \left(\frac{e^{\lambda_5 T} - e^{\lambda_5 t}}{e^{\lambda_8 T} - e^{\lambda_8 t}}\right).$$

4. Secondary growth equations[d]:

Normal growth stage $\left(\dfrac{^{206}Pb}{^{204}Pb}\right)_t$	Secondary growth

a) $\left(\dfrac{^{206}Pb}{^{204}Pb}\right)_{t'} = \left(\dfrac{^{206}Pb}{^{204}Pb}\right)_T + \left(\dfrac{^{238}U}{^{204}Pb}\right)_n \cdot (e^{\lambda_8 T} - e^{\lambda_8 t}) + \left(\dfrac{^{238}U}{^{204}Pb}\right)_n' \cdot (e^{\lambda_8 t} - e^{\lambda_8 t'}).$

b) $\left(\dfrac{^{207}Pb}{^{204}Pb}\right)_{t'} = \left(\dfrac{^{207}Pb}{^{204}Pb}\right)_T + \left(\dfrac{^{238}U}{^{204}Pb}\right)_n \cdot (e^{\lambda_5 T} - e^{\lambda_5 t}) + \left(\dfrac{^{235}U}{^{204}Pb}\right)_n' \cdot (e^{\lambda_5 t} - e^{\lambda_5 t'}).$

c) $\left(\dfrac{^{208}Pb}{^{204}Pb}\right)_{t'} = \left(\dfrac{^{208}Pb}{^{204}Pb}\right)_T + \left(\dfrac{^{238}U}{^{204}Pb}\right)_n \cdot (e^{\lambda_2 T} - e^{\lambda_5 t}) + \left(\dfrac{^{232}Th}{^{204}Pb}\right)_n' \cdot (e^{\lambda_2 t} - e^{\lambda_2 t'}),$

where $\left(\dfrac{^{238}U}{^{204}Pb}\right)_n'' = \left(\dfrac{^{235}U}{^{204}Pb}\right)_n'' = \left(\dfrac{^{232}Th}{^{204}Pb}\right)_n'' = 0.$

5. Secondary isochron[e]:

$$\frac{\left(\dfrac{^{207}Pb}{^{204}Pb}\right)_{t'} - \left(\dfrac{^{207}Pb}{^{204}Pb}\right)_t}{\left(\dfrac{^{206}Pb}{^{204}Pb}\right)_{t'} - \left(\dfrac{^{206}Pb}{^{204}Pb}\right)_t} = \left(\frac{^{235}U}{^{238}U}\right)_n \cdot \left(\frac{e^{\lambda_5 t} - e^{\lambda_5 t'}}{e^{\lambda_8 t} - e^{\lambda_8 t'}}\right).$$

6. Higher order growth equations:

   a) For solutions assuming continuously changing values of $\left(\dfrac{^{238}U}{^{204}Pb}\right)_n$ see PATTERSON and TATSUMOTO (1964) and WASSERBURG (1966).

   b) For stepwise mixing of multiple order growth systems see RUSSELL et al. (1966).

---

[c] For physical example see *Meteorites* (Fig. 12).

[d] There is no known physical example, but for a thorough discussion of how secondary growth might be applied to existing data see GAST (1969).

[e] Among the many physical examples see galenas of the Dominion Reef by BURGER et al. (1962) (see Fig. 14); and Bingham Canyon ores (STACEY et al. (1968) (see Fig. 15).

Table 11 (continued)

7. Higher order isochrons:

There is no rigorous higher order isochron or multistage isochron inasmuch as the loci of points do not all lie on one line. See KANASAWICH (1968a) for discussion. Some special cases of three-stage lead growth are amenable to mathematical solution (KANASEWICH and SLAWSON, 1964).

8. Constants:

For decay constants and value of $(^{238}U/^{235}U)_n$ see Table 1. For isotope ratios on primordial lead see Table 15.

9. Concordia diagram:

Developed by WETHERILL (1956a, b) particularly to solve discordant $^{207}Pb/^{235}U$ and $^{206}Pb/^{238}U$ ages. U and Pb contents must be known as well as lead isotopic composition. More recently the construction of concondia diagrams has been applied by STARIK et al. (1962b) to tektites and by ULRYCH (1967) to volcanic rocks:

a)
$$\frac{\left(\frac{^{206}Pb}{^{204}Pb}\right)_{t'} - \left(\frac{^{206}Pb}{^{204}Pb}\right)_T}{\left(\frac{^{238}U}{^{204}Pb}\right)_n'} = \frac{\left(\frac{^{238}U}{^{204}Pb}\right)_n}{\left(\frac{^{238}U}{^{204}Pb}\right)_n'} \cdot (e^{\lambda_8 T} - e^{\lambda_8 t}) + (e^{\lambda_8 t} - 1).$$

b)
$$\frac{\left(\frac{^{207}Pb}{^{204}Pb}\right)_{t'=0} - \left(\frac{^{207}Pb}{^{204}Pb}\right)_T}{\left(\frac{^{235}U}{^{204}Pb}\right)_n'} = \frac{\left(\frac{^{238}U}{^{204}Pb}\right)_n}{\left(\frac{^{238}U}{^{204}Pb}\right)_n'} \cdot (e^{\lambda_5 T} - e^{\lambda_5 t}) + (e^{\lambda_5 t} - 1).$$

This is a general form of the primary growth equations (Eqs. 4a, b, c) where $\left(\frac{^{238}U}{^{204}Pb}\right)_n'$ is not zero. In practice $\left(\frac{^{238}U}{^{204}Pb}\right)_n'$ is assumed to be the observed value and $\left(\frac{^{206}Pb}{^{204}Pb}\right)_T$ and $\left(\frac{^{207}Pb}{^{204}Pb}\right)_T$ are the primordial values in the material analyzed.

Therefore the equations allow for one other event between the time of origin of the earth and formation of the sample as analyzed. The systematics of the equations are thoroughly discussed in WETHERILL'S papers (1956a, b; 1963). For a paper that has considered the systematics in terms of earth systems see OVERSBY and GAST (1968a). Mistakes in interpretations have entered into the conclusions of most papers using the concordia relations. For physical example where the two-stage concordia conditions seem to apply see the treatment of abyssal basalts in TATSUMOTO (1966b) and ULRYCH (1967).

Table 12. *Equivalent symbols in common use*

	Canadian convention	Swiss convention	Value (Reference)
$\left(\dfrac{^{206}\text{Pb}}{^{204}\text{Pb}}\right)_t$	X	α	$t=0$; 18.66(1) 18.51(2)
$\left(\dfrac{^{206}\text{Pb}}{^{204}\text{Pb}}\right)_T$	a_0	α_0	9.56(3) 9.54(4) 9.346(5)
$\left(\dfrac{^{238}\text{U}}{^{204}\text{Pb}}\right)_n$	$137.8\,V$	μ_0	8.99(1) 8.7 (2) 9.09(6)
$\left(\dfrac{^{207}\text{Pb}}{^{204}\text{Pb}}\right)_t$	y	β	$t=0$; 15.79(1) 15.72(2)
$\left(\dfrac{^{207}\text{Pb}}{^{204}\text{Pb}}\right)_T$	b_0	β_0	10.42(3) 10.27(4) 10.218(5)
$\left(\dfrac{^{235}\text{U}}{^{204}\text{Pb}}\right)_n$	V	$\dfrac{\mu_0}{137.8}$	
$\left(\dfrac{^{208}\text{Pb}}{^{204}\text{Pb}}\right)_t$	Z	γ	$t=0$; 39.06(1) 38.44(2)
$\left(\dfrac{^{208}\text{Pb}}{^{204}\text{Pb}}\right)_T$	c_0	γ_0	30.0 (1) 29.71(3) 29.46(4) 28.96(5)
$\left(\dfrac{^{232}\text{Th}}{^{204}\text{Pb}}\right)_n$	W	$\mu \cdot k$	35.55(1) 34.8 (2) 38.3 (6)
$\left(\dfrac{^{232}\text{Th}}{^{238}\text{U}}\right)$	$\dfrac{W}{137.8\,V}$	k	3.92(1) 4.0 (2) 3.8 (3) 4.21(6)

References: (1) KANASEWICH (1968a); (2) DOE (1962b); (3) MURTHY and PATTERSON (1962); (4) CHOW and PATTERSON (1961), corrected according to CHOW and PATTERSON (1962b); (5) OVERSBY (1970); (6) STACEY *et al.* (1969).

Sample Handling. Most papers devote little attention to sample handling prior to lead extraction and purification. Lead contamination is particularly likely in most rock-lead work because of the widespread use of white lead house paint and the fallout of lead tetramethyl aerosols from gasoline. Great care must be taken to avoid contamination.

Method of Lead Isolation. Some satisfactory and commonly used methods of lead isolation were given in Table 2.

Methods of Abundance Measurement. Mass spectrometry (Table 2) is the common form of analysis for lead isotope abundances. Samples suitable for mass spectrometer calibration are in Table 3. Neutron activation ($^{208}Pb/^{204}Pb$) and alpha particle activation ($^{206}Pb/^{204}Pb$ and $^{208}Pb/^{204}Pb$) show promise of getting some data on materials having very low lead contents.

2. Meteorites, the Moon and Tektites

Studies of lead isotopes in meteorites have been very fundamental to all of earth science; the first reliable age of meteorites was a lead isochron age (PATTERSON, 1955), and the accepted age of the earth stems from meteorite studies (PATTERSON, 1953; PATTERSON et al., 1955).

The isotopic composition of lead in some meteorites is given in Table 13 and plotted in Fig. 12. The selection of meteorites is taken from the evaluation of MURTHY and PATTERSON (1962). Also given are the newly recommended values determined by OVERSBY (1970) where the ratios have been corrected for mass spectrometer bias, and the attempt has been made to eliminate samples that have suffered terrestrial lead contamination. STACEY et al. (1969) have pointed out that there is little change in derived values such as age of the earth or primary $^{238}U/^{204}Pb$ if all samples are in absolute ratios. Care should therefore be taken to make sure that all samples are in absolute ratios when using the values of primordial lead recommended by OVERSBY (1970). Other papers of interest are by STARIK et al. (1962a), SOBOTOVICH et al. (1964), and MARSHALL (1968).

The line that fits the data on Fig. 12 is a straight line and is a physical example of a primary isochron (Eq. 3) indicating that the meteorites were formed 4,550 million years ago (PATTERSON, 1956). From the slope of the line in the plot of $^{208}Pb/^{204}Pb$ versus $^{206}Pb/^{204}Pb$, a value of $^{232}Th/^{238}U$ may be calculated as 3.8 (MURTHY and PATTERSON, 1962).

The very recent results (TATSUMOTO and ROSHOLT, 1970; SILVER, 1970) on the samples returned by Apollo 11 from the Sea of Tranquillity on the moon are most exciting. The lead isotope ratios

(Table 13) of these materials are very radiogenic. The values of $^{207}Pb/^{204}Pb$ of the moon materials are 96–590 which are much greater than those of meteorites and the earth. These great ratios for the moon materials signify depletion of lead relative to uranium a long time ago when ^{235}U was still abundant. These data on the fines and breccia from

Table 13. *Isotopic composition of lead in some meteorites and Apollo 11 moon samples*

Meteorite and source of data	Atomic ratios		
	$^{206}Pb/^{204}Pb$	$^{207}Pb/^{204}Pb$	$^{208}Pb/^{204}Pb$
Radiogenic lead in some stone meteorites			
Nuevo Laredo (PATTERSON, 1955)	50.28	34.86	67.97
Richardton (MARSHALL and HESS, 1960)	38.16	27.70	56.27
Beardsley	13.67	12.40	31.85
Saratov (STARIK et al., 1958a)	19.53	16.70	40.25
Elenovka	21.54	16.94	39.86
Isotopic composition of primordial lead			
Average of Canyon Diablo and Henbury (PATTERSON, 1955)	9.50	10.36	29.49
Average of repeated analyses of Canyon Diablo troilite (CHOW and PATTERSON, 1961)	9.61	10.39	29.87
Average of Canyon Diablo, Burgavli and Arros troilites (STARIK et al., 1961)	9.74	10.70	30.28
Sardis troilite (MURTHY, 1962)	9.37	10.22	29.19
Average primordial lead (MURTHY and PATTERSON, 1962)	9.56	10.42	29.71
Newly recommended value of primordial lead in absolute ratios (OVERSBY, 1970)	9.346	10.218	28.96

Apollo 11 moon samples (TATSUMOTO and ROSHOLT, 1970)

Sample No.	Type	Concentration[a]			Atomic ratios		
		Pb (ppm)	U (ppm)	Th (ppm)	$^{206}Pb/^{204}Pb$	$^{207}Pb/^{204}Pb$	$^{208}Pb/^{204}Pb$
10003	crystalline	0.51	0.268	1.029	423.9 ± 25	198.0 ± 13	448.9 ± 25
10017	crystalline	1.56	0.854	3.363	410.0 ± 12	191.9 ± 6	435.0 ± 12
10020	vesicular	0.37	0.202	0.694	288.7 ± 18	139.0 ± 9	289.7 ± 18
10050	crystalline	0.29	0.156	0.531	295.8 ± 25	147.0 ± 12	287.3 ± 25
10057	vesicular	1.68	0.865	3.415	1241.5 ± 60	590.1 ± 30	1281.3 ± 60
10071	vesicular	1.71	0.873	3.434	199.7 ± 5	95.8 ± 3	206.7 ± 5
10061	breccia	1.74	0.674	2.572	249.1 ± 5	163.2 ± 3	258.2 ± 5
10084	fine material	1.39	0.544	2.092	261.9 ± 5	171.0 ± 3	270.1 ± 5

[a] Estimated errors for the concentrations are better than 2% for Pb and 1% for U and Th.

the moon lie very near the extension of the primary isochron of meteorites and the calculated U/Pb and Th/Pb ages support the age signified by the isochron. If the fines and breccia do not contain lead that has primarily been added to the moon from some foreign source, then the moon must be close to 4,600-m.y.-old, similar to that for meteorites. It should be kept in mind that this age is not necessarily the age at which the fines and breccia were formed, but the age at which the observed values of U/Pb and Th/Pb were established. While formation of fines and breccia at some time subsequent to 4,600 m.y. ago without disruption of U/Pb and Th/Pb is surprising, it still must be considered possible.

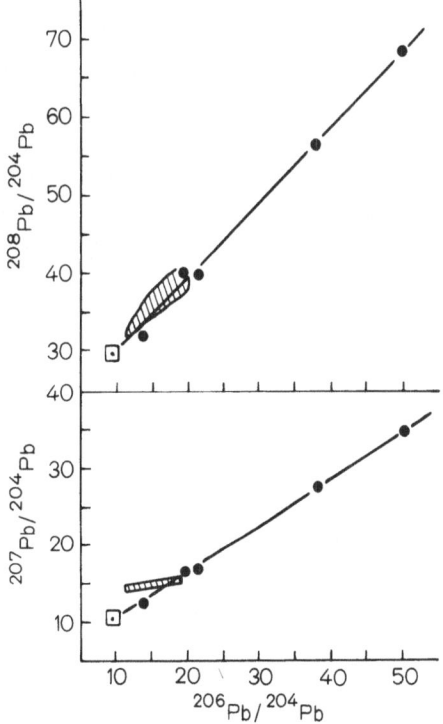

Fig. 12. A plot of ^{207}Pb/^{204}Pb and ^{208}Pb/^{204}Pb versus ^{206}Pb/^{204}Pb of troilite primordial lead ($\square$) and stone meteorite leads ($\cdot$) as selected by MURTHY and PATTERSON (1962). The solid line in the lower plot is a primary isochron with a slope of 0.59 and the slope of the solid line in the upper plot represents a Th/U of 3.8. Cross-hatched areas enclose terrestrial leads

The lead in the lunar volcanic rocks also give nearly concordant ages between about 3,700 and 4,200 m.y. The lead isotope ages therefore infer that the moon was active for about a billion years after its formation. The difference in the isotopic trends between leads in the

Table 14. *Isotopic composition of lead in tektites*

Sample	Concentrations (ppm)			Isotopic ratios (atomic)			References
	U	Th	Pb	$^{206}Pb/^{204}Pb$	$^{207}Pb/^{204}Pb$	$^{208}Pb/^{204}Pb$	
Philippite[a]	—	—	~5	18.83	15.70	39.07	TILTON (1958)
Philippite[b]	2.5	18	1.4	18.69	15.80	39.14	STARIK et al. (1963)
Australite[c]	1.7	9.2	2.9	18.81	15.60	38.70	TILTON (1958)
Moldavite[a]	—	—	~6	18.55	15.65	38.41	TILTON (1958)
Moldavite[b]	2.0	26	2.4	18.08	15.66	38.25	STARIK et al. (1963)
Moldavite[c]	2.1	—	6.2	18.8	15.8	39.5	WAMPLER et al. (1969)
Moldavite[c]	2.6	—	4.6	18.4	15.4	38.0	WAMPLER et al. (1969)
Moldavite[b]	2.3	—	3.7	20.42	16.05	39.45	STARIK et al. (1962b)
Indochinite I[b]	2.3	19	2.4	18.78	15.75	39.07	STARIK et al. (1963)
Indochinite II[b]	2.7	18	8.6	18.64	15.63	38.81	STARIK et al. (1963)
Indochinite[c]	1.9	—	2.8	18.3	15.3	38.2	WAMPLER et al. (1969)
Ivory Coast[c]	0.53–1.2	—	$\leq$1.3	17.5	14.8	35.6	WAMPLER et al. (1969)
Ivory Coast[c]	0.57	—	$\leq$1.4	18.1	15.3	36.7	WAMPLER et al. (1969)
Bediasite[c]	—	—	—	18.3	15.3	37.5	WAMPLER et al. (1969)
Georgia[c]	—	—	—	17.8	14.8	36.4	WAMPLER et al. (1969)

[a] Concentration by volume of dithizone turned red ($\pm$ 20%).
[b] Th determined colorimetrically; method used for U and Pb is not known.
[c] Concentration by isotope dilution.

fines and breccia on the one hand and the volcanic rocks on the other also shows that the fines and breccia are not simply regolith of the underlying volcanics. Much of the fines and breccia must have been transported to their present location. How representative the Apollo 11 data are of the moon as a whole will shortly be tested by results on samples returned by Apollo 12. Additional speculation therefore does not seem to be warrented at this time.

Meteorites have large variations in values of $^{207}Pb/^{204}Pb$ – from 12 to 35 – and those of the moon are even greater – 96–590 – whereas for the earth this ratio is rather uniform, ranging mainly between 15 and 16. TILTON (1958) used the difference in the nature of $^{207}Pb/^{204}Pb$ between meteorites and the earth to try to determine whether tektites are terrestrial or extraterrestrial in origin. All tektites (Table 14) are found to have a terrestrial type of $^{207}Pb/^{204}Pb$ value (such as included in the cross-hatched area on $^{207}Pb/^{204}Pb$ of Fig. 12). Thus arguments based on lead isotopes suggest a derivation of tektites from a terrestrial type of source rather than a meteorite or moon source. WAMPLER et al. (1969), through analyses of tektites and rocks of possible impact crater sources, have shown that the lead isotopes support the growing body of data relating the Ivory Coast tektites to the Bosumtwi crater. They also show that moldavites are compatible with a derivation from Ries crater and suggest that other tektites may be ultimately derived from pelagic sediments.

3. Isotopes in Nature

Age of Earth. All leads from terrestrial rocks except those from uranium and thorium ores are enclosed in the cross-hatched areas of Fig. 12. If the earth and meteorites were totally unrelated in time of origin or in the isotopic composition of primordial lead, there is no reason why terrestrial lead should cluster about the primary isochron for meteorites. The proximity of terrestrial leads to the primary meteorite isochron has been used by PATTERSON (PATTERSON et al., 1955; PATTERSON, 1956; MURTHY and PATTERSON, 1962) in estimating the age of the earth as well as the isotopic composition of terrestrial primordial lead. Little change has been made in the estimate since PATTERSON's first work; nevertheless, detailed studies on terrestrial materials do indicate that the earth may be as much as 200-m.y.-older (TILTON and STEIGER, 1965, 1969): however, if the earth is even that much older, either there must have also been differentiation of U relative to lead in the early history (PATTERSON and TATSUMOTO, 1964) or the values of primordial lead for the earth and for meteorites must differ (DOE et al., 1965).

Table 15. *Least radiogenic leads*

Locality	$^{206}Pb/^{204}Pb$	$^{207}Pb/^{204}Pb$	$^{208}Pb/^{204}Pb$	References
	Galenas ($^{206}Pb/^{204}Pb < 13$; $^{208}Pb/^{204}Pb < 33$)			
South Africa				
Rosetta	present day	12.58	14.18	32.59 ⎫
Daylight	present day	12.62	14.21	32.77 ⎬ ULRICH *et al.* (1967)
Consort	present day	12.52	14.22	32.86 ⎬
French Bobs	present day	12.51	14.21	32.89 ⎭
French Bobs	present day	12.461[a]	14.077[a]	32.285[a] STACEY *et al.* (1969)
North Congo Shield, Africa				
Kokosho	present day	12.75	14.39	32.62 ⎫ RUSSELL and FARQUHAR (1960);
		12.73	14.32	32.65 ⎬ EBERHARDT *et al.* (1955)
Kokosho II	present day	12.63	14.32	32.71 EBERHARDT *et al.* (1962)
	Feldspars ($^{206}Pb/^{204}Pb < 14$; $^{208}Pb/^{204}Pb$ corrected < 34)			
Ukrainian Shield, Europe				
	present day	13.80	14.55	33.20 ⎫ TUGARINOV *et al.* (1963a)
	corrected 3200 m.y.	13.00	14.46	32.40 ⎭
Baltic Shield, Europe				
	present day	13.73	14.60	33.80 ⎫ TUGARINOV *et al.* (1963a)
	corrected 2600 m.y.	13.65	14.59	33.50 ⎭

Table 15 (continued)

Locality		$^{206}Pb/^{204}Pb$	$^{207}Pb/^{204}Pb$	$^{208}Pb/^{204}Pb$	References
Canadian Shield, North America					
Vermilion Granite, Minnesota and Ontario	present day[b]	13.63	14.60	33.34 ⎫	Doe *et al.* (1965)
	present day[c]	13.82	14.66	33.57 ⎬	
	corrected 2600 m.y.[c]	13.52	14.61	33.37 ⎭	
Manitouwadge Lake, Ontario	present day[c]	13.63	14.66	33.50	Tilton and Steiger (1965)
Rocky Mountain region, North America					
Quartz monzonite, Wyoming	present day	13.68	14.89	33.71 ⎫	Heimlich and Banks (1968)
	corrected 2900 m.y.	13.57	14.87	— ⎬	
Pegmatite, Wyoming	present day[d]	13.84	15.02	34.00	Catanzaro and Gast (1960)
Pyrite					
Canadian Shield, North America					
Pine Cone Lake, Ontario	present day[e]	13.64	14.75	33.87	Wampler and Kulp (1964)
Whole Rocks					
Aldan Shield, Asia					
Lake Baikal	present day[b]	13.84	14.86	36.30	Sobotovich *et al.* (1965)

[a] Absolute ratio. — [b] No U, Th, or Pb concentrations available. — [c] Average of three samples. — [d] Pegmatite sample containing 100 ppm Pb. — [e] Corrections for decay of U and Th are insignificant. Average of two analyses.

A second method of estimating a minimum age of the earth is to hunt for the last radiogenic leads (Table 15). A number of galenas have $^{206}Pb/^{204}Pb < 13$ and $^{208}Pb/^{204}Pb < 33$. In terms of a primary isochron these galenas would be about 3,400–3,500-m.y.-old and 3,100 m.y. would be the age calculated from $^{208}Pb/^{204}Pb$. No direct evidence of rock leads that are as unradiogenic as galenas is known; however, a number of feldspars, a pyrite and a whole rock have $^{206}Pb/^{204}Pb < 14$ and $^{208}Pb/^{204}Pb < 34$.

That there is hope of finding terrestrial rocks somewhat older than 3,500 m.y. is implied in the secondary isochron study of igneous rocks about 1,900-m.y.-old from the Churchill province in Canada (DOE, 1967). The slope of leads from three feldspars ($\Delta^{207}Pb/^{204}Pb)/(\Delta^{206}Pb/^{204}Pb)$ is 0.521; the source age (t) may then be $\geq 3,800$ m.y.

Primary or Single Stage Growth Curves. A physical example of a close approach to primary growth curves (Eq. 2a, b, c) is found for the lead isotope data of *conformable ore deposits* (STANTON and RUSSEL, 1959).

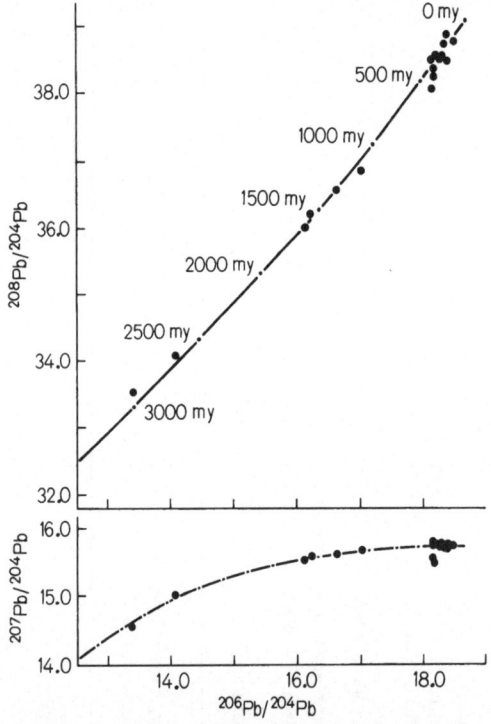

Fig. 13. $^{207}Pb/^{204}Pb$ and $^{208}Pb/^{204}Pb$ versus $^{206}Pb/^{204}Pb$ for stratiform ores which appear to have formed at approximately the same time as the enclosing sediments and approach conditions of single-stage lead isotope growth. Names and data on the deposits are given in Table 16. Parameters of the growth curves are those of KANASEWICH (1968c) and are given in Tables 12 and 1

Table 16. *Data on stratiform ores whose lead approaches single stage lead isotope development*

Location	Host rock	No. of samples in average	$^{206}Pb/^{204}Pb$	$^{207}Pb/^{204}Pb$	$^{208}Pb/^{204}Pb$	Ref.
Bleiberg, Austria	Triassic limestones	2	18.41	15.80	38.46	3
	Triassic limestones	4	18.52	15.83	38.87	7
Mibladen, Morocco	Mesozoic limestones	3	18.34	15.74	38.72	1
Halls Peak, New South Wales, Australia	Greywackes and carbonaceous rocks, Upper Permian(?)	6	18.491	15.754	38.770	2
Kupferschiefer Marl-Slate Northern Pennines, Scotland	Middle Permian rocks	1	18.20	15.54	38.15	5
Mansfeld, East Germany	Middle Permian rocks	9	18.17	15.80	38.37	6
Spremberg, East Germany	Middle Permian rocks	11	18.28	15.75	38.50	6
Meggen, West Germany	Middle and Upper Devonian slates	1	18.37	15.70	38.60	8
	Middle and Upper Devonian slates	1	18.42	15.74	38.62	7
Rammelsberg, West Germany	Middle Devonian slates	2	18.15	15.58	38.21	8
Cobar, New South Wales, Australia	Upper Silurian sandy and shaly rocks	8	18.210	15.772	38.548	2
Captains Flat, New South Wales, Australia	Carbon- and carbonate-bearing shales of Early Silurian(?)	7	18.178	15.766	38.565	2
Bathurst, New Brunswick, Canada	Middle Ordovician metamorphosed rocks	3	18.291	15.781	38.526	2
	Middle Ordovician metamorphosed rocks	1	18.20	15.69	38.24	4
	Middle Ordovician metamorphosed rocks	1	18.204[a]	15.655[a]	38.122[a]	14
Read Roseberry, Tasmania, Australia	Cambrian tuffaceous shale, highly folded	7	18.374	15.746	38.469	2
Balmat, New York	Marbles about 1100 m.y. in age	1	17.019	15.677	36.857	10
	Marbles about 1100 m.y. in age	4	16.96	15.54	36.57	9
	Marbles about 1100 m.y. in age	1	16.935[a]	15.505[a]	36.423[a]	14

[a] Absolute ratios.

Table 16 (continued)

Location	Host rock	No. of samples in average	$^{206}Pb/^{204}Pb$	$^{207}Pb/^{204}Pb$	$^{208}Pb/^{204}Pb$	Ref.
Sullivan, British Columbia, Canada	Belt Supergroup rocks (1000—1800 m.y. in age)	2	16.64	15.63	36.63	3
	Belt Supergroup rocks (1000—1800 m.y. in age)	2	16.63	15.64	36.58	11
Mt. Isa, Australia	Metamorphosed rocks approximately 1600—1700-m.y.-old	5	16.22	15.60	36.24	2, 12
Broken Hill, Australia	Metamorphosed rocks approximately 1600—1700-m.y.-old	6	16.12	15.54	36.06	12
	Metamorphosed rocks approximately 1600—1700-m.y.-old	1	16.07	15.52	36.03	4
	Metamorphosed rocks approximately 1600—1700-m.y.-old	1	16.007[a]	15.397[a]	35.675[a]	14
Geneva Lake, Ontario, Canada	Precambrian metamorphosed rocks	2	14.08	15.02	34.10	3
	Precambrian metamorphosed rocks	1	14.002[a]	14.870[a]	33.716[a]	14
Manitouwadge Lake, Ontario, Canada	Archean metamorphosed rocks approximately 2700—2800-m.y.-old	2	13.34	14.56	33.27	3
	Archean metamorphosed rocks approximately 2700—2800-m.y.-old	4	13.400	14.565	33.545	2
	Archean metamorphosed rocks approximately 2700—2800-m.y.-old	1	13.25	14.42	33.18	13
	Archean metamorphosed rocks approximately 2700—2800-m.y.-old	1	13.211[a]	14.401[a]	33.069[a]	14

References: (1) CAHEN et al. (1958); (2) OSTIC et al. (1967); (3) RUSSELL and FARQUHAR (1960); (4) DOE (1962b); (5) MOORBATH (1962); (6) KAUTSCH et al. (1964); (7) EBERHARDT et al. (1955); (8) GEISS (1954); (9) DOE (1962a); (10) REYNOLDS and RUSSELL (1968); (11) SINCLAIR (1966); (12) KOLLAR et al. (1960); (13) DOE et al. (1965); (14) STACEY et al. (1969).

[a] Absolute ratios.

These ore deposits are tabular (grossly conformable with the country rock), have ill-defined margins, and are pyrite-rich; many are in rocks bearing organic matter and most are in metamorphic terranes. Stanton has interpreted the terrane containing the deposits to be of the island-arc and related tectonic classes. The data for some of these deposits are given in Table 16 and are shown in Fig. 13. The source of the lead in such deposits is of uncertain origin (see Section 9, Ore Genesis).

Secondary or Two Stage Growth Curves. Many examples are now known which illustrate secondary isochrons. One particularly good study deals with the sulfide mineralization of the Dominion Reef (BURGER et al., 1962). Samples of whole rock and of detrital monazite from the Dominion Reef conglomerates gave ages close to 3,100 m.y. (NICOLAYSEN et al., 1962). If this figure indicates the age of the source of the lead in galenas, then the galena mineralization occurred 2,000–2,240 m.y. ago, according to a slope of 0.39 for the line (Fig. 14).

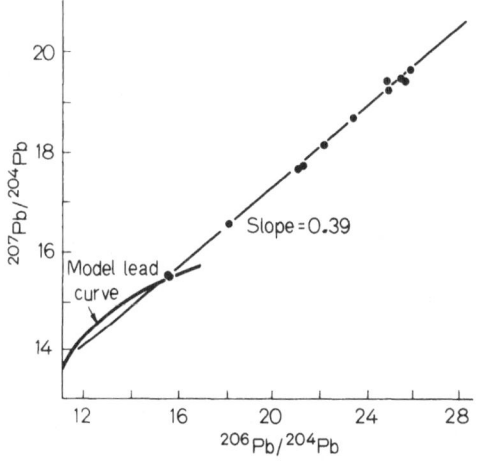

Fig. 14. Lead isotope plot of galenas from the Dominion Reef (after BURGER et al., 1962). Interpretation of the slope is given in the text

The mineralization age is close to that of widespread igneous activity in South Africa, such as intrusion of the Bushveld Complex and metamorphism of the Witwatersrand system, at 1,950 m.y. (NICOLAYSEN et al., 1958). The lead isotopic data support the petrologic evidence presented by LIEBENBERG (1955) that the lead was largely derived from detrital uraninite in the Dominion Reef conglomerate.

Because the development of lead isotope ratios has occurred in two or more stages, the observed ratios may be greater than they would be in a single stage development because the lead has evolved in an integrated system with $^{238}U/^{204}Pb$ > about 9. Such lead is *radiogenic* in comparison with a single stage lead of that age. Special radiogenic leads having future model lead ages are sometimes referred to as J-type or Joplin type. If the lead isotope ratios are less than they would be in a single stage development because of an integrated $^{238}U/^{204}Pb$ < about 9, the term *unradiogenic lead* relative to a single stage lead of that age will be used. Unradiogenic leads are sometimes called B-type or Bleiberg type. Leads from the Bleiberg deposits do not fit the isotopic definition, however, and the use of the term B-type should be discontinued.

4. Observed Values of $^{238}U/^{204}Pb$ and $^{232}Th/^{204}Pb$ in Rocks

The lead isotopic composition of igneous rocks and ores places constraints on the observed atomic values of $^{238}U/^{204}Pb$ and $^{232}Th/^{204}Pb$ in the source rocks. Knowledge of the values of these ratios observed in various rock types is therefore desirable in radiogenic tracer studies (Fig. 15). References to the data in the figure are given in Table 17 or in Section 5, Whole-Rock Studies, Precambrian and Paleozoic. In Fig. 15, Group 1 is the mafic and ultramafic inclusions of LOVERING and TATSUMOTO (1968) from pipes Mesozoic or Cenozoic in age and a gabbro inclusion from Ichinomugata, Japan (HEDGE and KNIGHT, 1969). Group 2 includes abyssal basalts, tholeiites, olivine tholeiites, tholeiites with alkalic affinities, and calcalkaline basalts that are Mesozoic or Cenozoic in age. Petrographic descriptions of many of these samples are not available, but where the studies report xenocrysts in these basalts, they are grouped into a separate category (Group 4). Subalkaline basalts form the major group of igneous rocks which have numerous values of $^{238}U/^{204}Pb$ and $^{232}Th/^{204}Pb$ less than 9.0 and 35 respectively. Such low values would, in time, generate unradiogenic uranium and thorium derived leads along secondary isochrons. Some, such as abyssal basalts, seem to have values of $^{238}U/^{204}Pb$ that are undifferentiated in the magmatic process (TATSUMOTO, 1966b); and reverse differentiation (i.e., decrease in $^{238}U/^{204}Pb$ in the magmatic process) is known such as in the Type I basalts of WELKE et al. (1968) from Iceland. Apparent reverse differentiation is common for $^{232}Th/^{204}Pb$ and has been ascribed to partial melting by TATSUMOTO (1966b). The Group 3 includes alkali basalt, melilites, nepheline basalts, trachybasalts, nephelinites, and Kimberlite that are Mesozoic and

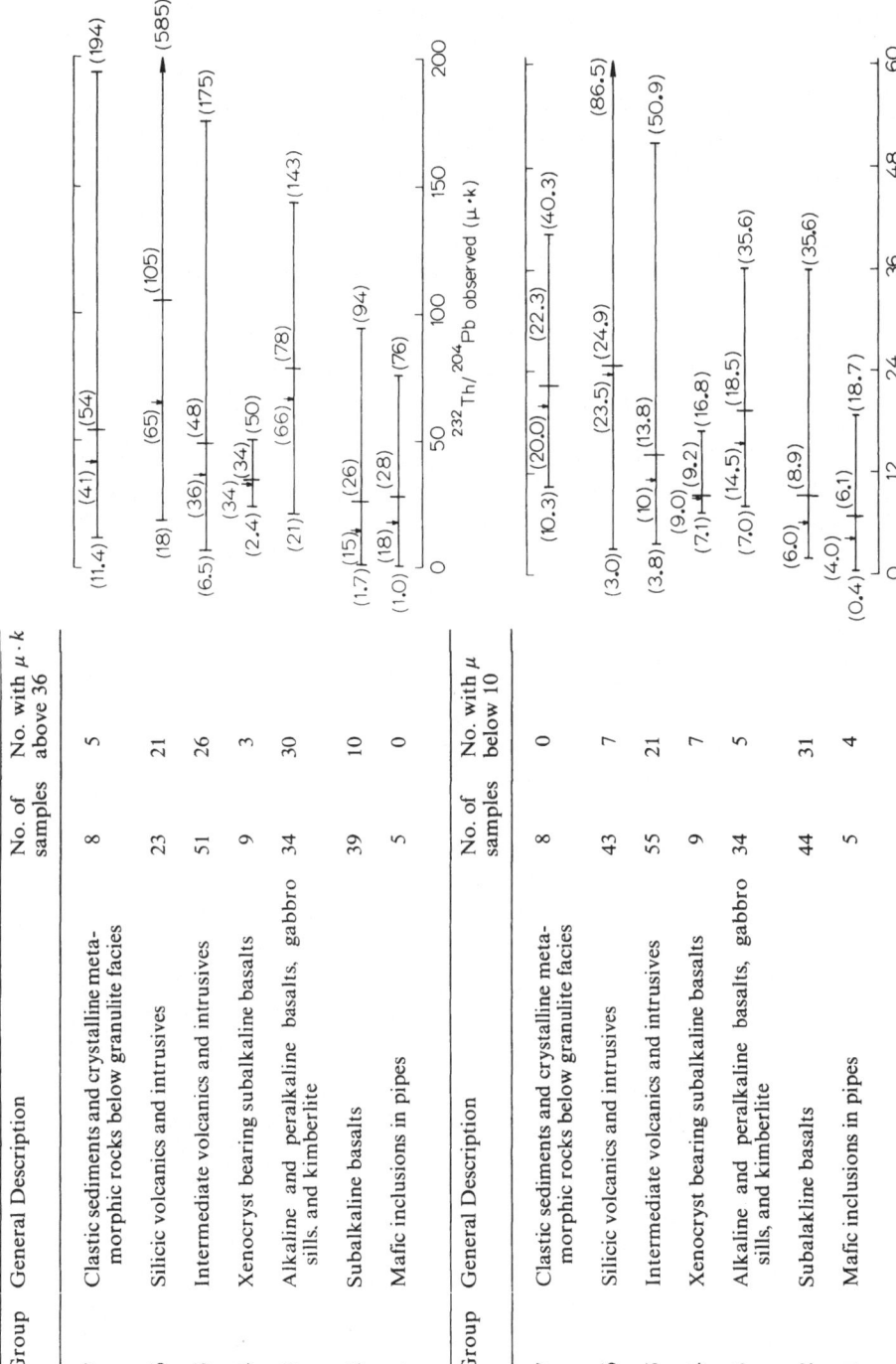

Fig. 15. Extremes, means (long vertical bar), and fiftieth percentiles (arrows) for various rock types for which lead isotope data are available. Numbers in parentheses give values for each indicator (References to data are given in Section III, part 5 — Whole rock studies, Precambrian and Paleozoic — and in Table 17)

Group	General Description	No. of samples	No. with $\mu \cdot k$ above 36
7	Clastic sediments and crystalline metamorphic rocks below granulite facies	8	5
6	Silicic volcanics and intrusives	23	21
5	Intermediate volcanics and intrusives	51	26
4	Xenocryst bearing subalkaline basalts	9	3
3	Alkaline and peralkaline basalts, gabbro sills, and kimberlite	34	30
2	Subalkaline basalts	39	10
1	Mafic inclusions in pipes	5	0

Group	General Description	No. of samples	No. with μ below 10
7	Clastic sediments and crystalline metamorphic rocks below granulite facies	8	0
6	Silicic volcanics and intrusives	43	7
5	Intermediate volcanics and intrusives	55	21
4	Xenocryst bearing subalkaline basalts	9	7
3	Alkaline and peralkaline basalts, gabbro sills, and kimberlite	34	5
2	Subalkaline basalts	44	31
1	Mafic inclusions in pipes	5	4

Cenozoic in age. $^{238}U/^{204}Pb$ and $^{232}Th/^{204}Pb$ are generally greater for the alkalic group than for the subalkalic group. The ratios are nearly always increased in the differentiation process. The Group 5 of intermediate volcanics and intrusives includes andesites, trachyandesites, trachytes, andesineandesite and Hawaiite, mugearites, dacite, diorites and quartz diorites of all ages; however, data are available for only one sample of quartz diorite (ZARTMAN, 1965). As in the subalkalic basalts, $^{238}U/^{204}Pb$ is often unchanged or decreased in the magmatic process (PETERMAN et al., 1969) as is $^{232}Th/^{204}Pb$, although strong normal differentiation is also known such as for the intermediate compositions from Iceland of WELKE et al. (1968).

The silicic grouping (6) includes rhyolites, obsidians, rhyodacites, granophyres, granodiorites, quartz monzonites, and granites of all ages. Only Precambrian rocks are included where the observed $^{238}U/^{204}Pb$ or $^{232}Th/^{204}Pb$ is in approximate agreement with that needed to grow the observed lead isotopic composition from a model initial lead. This precaution is necessary because the value of the observed ratios often appears to have been recently disturbed by incipient weathering or leaching of some sort. ZARTMAN (1965) and ROSHOLT and PETERMAN (1969) report examples of recent uranium depletion relative to lead although the $^{232}Th/^{204}Pb$ appeared to be undisturbed as reported by ROSHOLT and PETERMAN (1969); the SOBOTOVICH et al. (1963a) report contains concentration data that suggest recent uranium enrichment relative to lead (Table 10). The ratio values range widely and independently of rock type within this grouping. The observed values of these ratios are often great even when the contained lead is unradiogenic as in unaltered Cenozoic silicic igneous rocks (DOE, 1967).

All isotopically analyzed sediments and crystalline metamorphic rocks below granulite facies have large observed values of $^{238}U/^{204}Pb$ and most have great values for $^{232}Th/^{204}Pb$; however, many low values are anticipated because there are many unradiogenic leads reported for Precambrian crystalline rocks in the heart of shield areas (see Fig. 17) on which uranium and lead concentration data are not reported but which would give values calculated from the lead isotopic composition of $^{238}U/^{204}Pb$ less than 9 and $^{232}Th/^{204}Pb$ less than 35. Also the uranium and lead concentration data in LAMBERT and HEIER (1968) may be used to calculate values of $^{238}U/^{204}Pb$ by assuming a value of 1.4 (± 0.05) percent ^{204}Pb in the reported lead concentration. The calculated $^{238}U/^{204}Pb$ will probably not have a calculation error greater than 3.5 percent. The calculated values for the amphibolite facies rocks as well as for the granulite facies rocks are much below 9, as has been pointed out elsewhere (DOE et al., 1968).

5. Whole-Rock Studies, Precambrian and Paleozoic

Knowledge of the isotopic composition of lead in Precambrian and Paleozoic upper crustal rocks is important in answering many questions of genesis, such as the source of igneous rocks and ores (for a compilation lation of data see Appendix B). Fig. 16 shows a wide spectrum of present day isotopic compositions in these rocks. All North American data are enclosed in the area shown; these data are from PATTERSON (1953)*, PETERMAN et al. (1967)*, DOE et al. (1967)*, ZARTMAN (1965)*, and

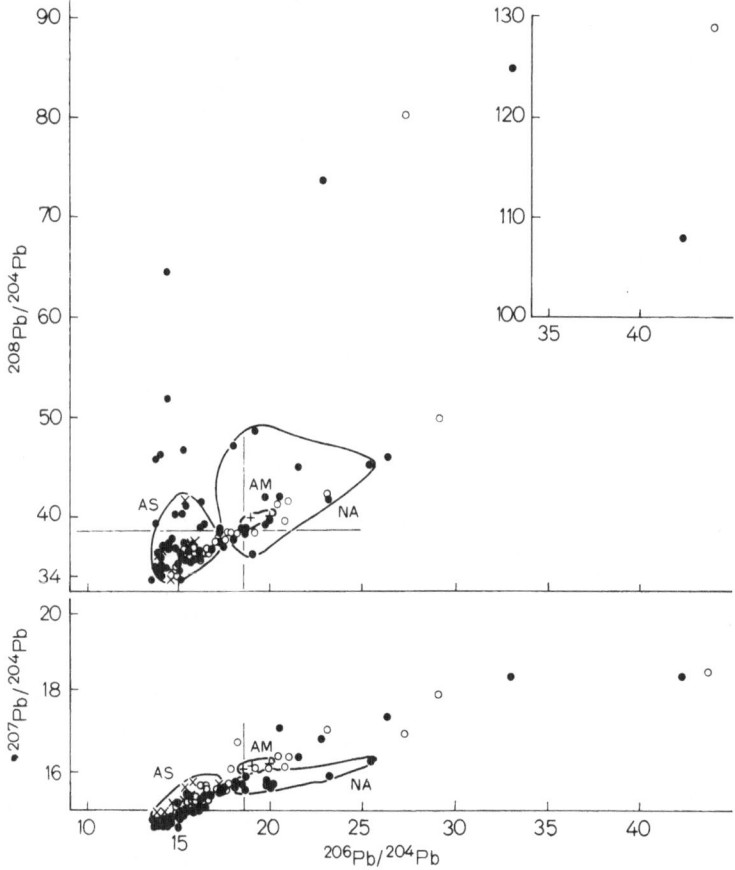

Fig. 16. $^{207}Pb/^{204}Pb$ and $^{208}Pb/^{204}Pb$ versus $^{206}Pb/^{204}Pb$ for whole-rock samples of Precambrian and Paleozoic crystalline rocks and composites: ○ Ukrainian Shield; ● Baltic Shield and Scotland; × Aldan Shield (AS); + Altai Mountains (AM); ●North America (NA) (References to data are given in Section III, part 5 — Whole rock studies, Precambrian and Paleozoic)

---

* Concentration data from these papers are included on Fig. 15.

MURTHY and PATTERSON (1961)*. (All decompositions were by fusions or HF-HClO$_4$ with concentrations checked by fusions.) The average minus the two most radiogenic samples is 19.14 for $^{206}Pb/^{204}Pb$, 15.55 for $^{207}Pb/^{204}Pb$, and 39.89 for $^{208}Pb/^{204}Pb$. Lines are drawn on the diagram to show the approximate position of a terrestrial lead that lies on the meteorite geochron (a model lead of zero age). The rocks of the Altai Mountains in Asia (SOBOTOVICH and GRASHCHENKO, 1965) also appear to be radiogenic.

Such radiogenic rocks are not necessarily characteristic for some large segments of the earth's crust as may be seen (Fig. 16) from the data on the Precambrian crystalline rocks of the Baikalian block of the Aldan Shield in Asia (SOBOTOVICH et al., 1965). (The Aldan Shield rocks were decomposed in HF-COOH$_2$, which may not decompose some radio-active minerals such as zircon; however, other studies have shown that only minor amounts of lead in the whole rock are held in such phases.)

Data from the Baltic and Ukrainian Shield areas range widely, which makes averages difficult to determine. The leads in rocks of the Ukrainian Shield (SOBOTOVICH et al., 1963a, b, c) are also probably unradiogenic today, as the isotope ratios minus the two most radiogenic samples are: 17.61 for $^{206}Pb/^{204}Pb$, 15.66 for $^{207}Pb/^{204}Pb$, and 37.80 for $^{208}Pb/^{204}Pb$. For a selection of data from the Ukrainian Shield see Table 10. The Baltic Shield is more problematical although many of the rocks are unradiogenic (data on the Baltic Shield are from VINOGRADOV and ZYKOV, 1955*; STARIK and SOBOTOVICH, 1956; ZHIROV and ZYKOV, 1956; and SOBOTOVICH et al., 1963c). Even after the three most radio-genic samples are excluded, $^{208}Pb/^{204}Pb$ is still radiogenic: 17.69 for $^{206}Pb/^{204}Pb$, 15.70 for $^{207}Pb/^{204}Pb$, and 39.14 for $^{208}Pb/^{204}Pb$.

A detailed study of high grade Precambrian metamorphic rocks from the Lewisian basement of northwest Scotland has recently become available (MOORBATH et al., 1969). As the leads were extracted by volatili-zation, there is some danger that not all the lead was extracted; however, other studies indicate that volatilized leads from old rocks would be more radiogenic, if anything, than the whole-rock values in this case. (The authors feel they probably did get complete extraction as the temperature of volatilization was 1,200° C rather than the usual 1,050° C.) Even so, all present day values of $^{206}Pb/^{204}Pb$ are strongly nonradiogenic – in the range 13.5 to 16.5 – regardless of rock type. Rocks of granulite facies metamorphism about 2,600 m.y. ago and those in the transition zone to an area of amphibolite facies metamorphism about 1,400–1,600 m.y. ago also have nonradiogenic values of $^{208}Pb/^{204}Pb$ – in the range 33.9 to 37.5. This range also includes many samples from the region of

---

* Concentration data from these papers are included on Fig. 15.

subsequent amphibolite facies metamorphism; however, in 40 percent of these samples, $^{208}Pb/^{204}Pb$ is greater than 40 and must be considered radiogenic if the ratios are close to equal of those for the whole-rocks. Significant thorium but apparently little uranium may have been added during the amphibolite facies metamorphism to account for these isotopic relationships.

Where the highest grade of metamorphism in a region is of the amphibolite facies (such as for the Precambrian rocks analyzed from North America), present-day values of $^{206}Pb/^{204}Pb$ and $^{208}Pb/^{204}Pb$ are generally radiogenic. The fact that the granulite facies rocks studied by MOORBATH et al. (1969) are nonradiogenic give some isotopic support to the hypotheses that granulite (DOE et al., 1968) and higher rank metamorphism (DOE, 1968) would materially reduce the values of U/Pb and Th/Pb in a rock, a factor of great importance to igneous petrology.

6. Cenozoic Sediments, Whole-Rock and HCl-Soluble Lead

Lead isotopic composition of the main sediment types (for a compilation see Appendix C) is more a function of age of source areas for the sediment than of rock type because the source rocks contain significant quantities of U and Th relative to lead (Fig. 17). Young sediments in restricted basins draining Precambrian terranes, such as Hudson Bay, Great Slave Lake, Great Bear Lake, Lake Superior, and the Baltic Sea (CHOW, 1965; HART and TILTON, 1966), tend to be highly radiogenic and variable in lead isotope ratios.

On the other hand, sediments, well-mixed because they are far from their source, tend to be isotopically rather uniform even if the detritus is Precambrian in age, but the lead is still somewhat radiogenic (DOE et al., 1966). The isotopic composition in the insoluble residues of these sediments, all roughly of the same sedimentation age, appears to be more a function of grain size but which may in reality be more a function of K-feldspar content. The coarse fractions are less radiogenic than the silt- or clay-sized fractions, and the lead isotope ratios still reflect the Precambrian ancestry of the detritus (MUFFLER and DOE, 1968). The finer fractions of detritus approach more closely the isotopic composition of the HCl-soluble fraction.

Pelagic sediments from the ocean basins (CHOW and PATTERSON, 1962a; CHOW and TATSUMOTO, 1964; WAMPLER and KULP, 1964; CHOW, 1968b) fall in restricted ranges of lead isotopic composition (Fig. 17), although the means from the Atlantic, Indian, and Antarctic Oceans and the Red Sea are somewhat more radiogenic than those of the Pacific Ocean. CHOW and PATTERSON (1962a) have concluded that this relationship reflects a source terrane around the Pacific which is younger than the terranes which surround the other areas.

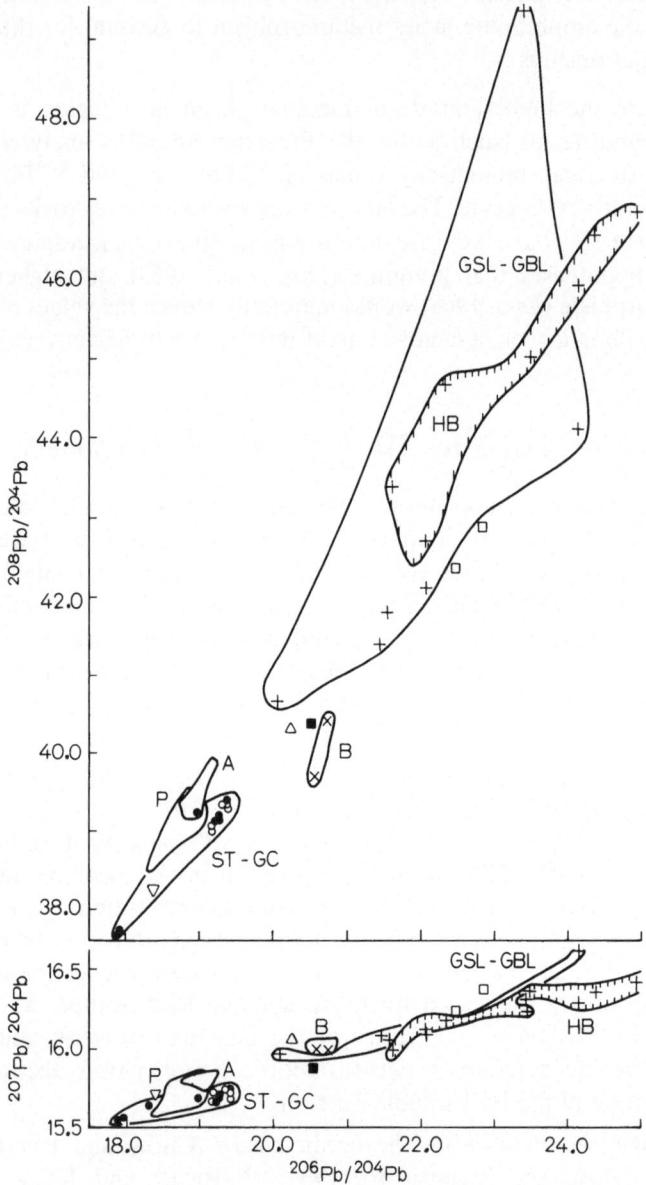

Fig. 17. $^{207}Pb/^{204}Pb$ and $^{208}Pb/^{204}Pb$ versus $^{206}Pb/^{204}Pb$ for sediments of Cenozoic age from various localities. (References to data are given in Section III, part 6 — Cenozoic sediments whole-rock and HCl soluble lead):

Area or Symbol	Description
A	Pelagic sediments of the Red Sea and basins of the Atlantic, Antarctic, and Indian Oceans except for the Gulf of Aqaba, HCl soluble lead
△	Sediment from the Gulf of Aqaba, HCl soluble lead
P	Pelagic sediments from the Pacific Ocean basin, HCl soluble lead
ST-GC	Calcareous clastic sediments from the Salton Trough and a manganese nodule from the Gulf of California:
○	HCl soluble lead
●	residue lead
B	Sediments from the Baltic Sea:
×	HCl soluble lead
	Sediments from Lake Superior in the Canadian Shield:
☐	HCl and water soluble lead
■	residue lead
HB	Sediments from Hudsons Bay in the Canadian Shield, HCl soluble lead
GSL-GBL	Sediments from Great Bear Lake and Great Slave Lake in the Canadian Shield:
+	HCl soluble lead
▽	Sediments from the Mediterranian Sea, HCl soluble lead

7. Cenozoic and Mesozoic Igneous Rocks

Ocean Basins. The most fundamental rocks upon which to obtain isotopic data are currently the abyssal basalts from oceanic ridges and rises (TATSUMOTO, 1966b) and basalts from islands in "open ocean" structures such as Hawaii (references in Table 17). The data on abyssal basalts which are given in Table 18 are shown in enclosed areas along with all Hawaiian data in Fig. 18. The isotope ratios are clearly not uniform, although the average of all samples would approach a modern lead within the limits of error that such models may be calculated. The age of origin for the source of abyssal basalts may be quite old, on the order of 1,000 m.y., and there may be little differentiation of uranium relative to lead as TATSUMOTO has demonstrated (also see ULRYCH, 1967). TATSUMOTO has also shown that the Th/Pb is apparently equal to or less than that in the source.

Island volcanics from oceanic ridges and rises have lead isotope ratios (references in Table 17; data on Fig. 18 with selected data on Table 18) which are in general more radiogenic than those of abyssal basalts and Hawaiian volcanics (DOE, 1968a). This relationship is

Table 17. *Sources of isotopic and concentration data on Cenozoic and Mesozoic igneous rocks*

Region	Sources of isotopic and concentration data
Atlantic Ocean basin	
Abyssal basalts (Fig. 18–20)	TATSUMOTO (1966b)[a]
Island volcanics, Mid-Atlantic Ridge (Fig. 18)	GAST et al. (1964) – ASCENSION and GOUGH; GAST (1967, 1969) – St. Helena; WELKE et al. (1968)[a] – Iceland; OVERSBY and GAST (1968b)[a] – TRISTAN DA CUNHA and FAIAL, Azores
Island volcanics, other localities (Fig. 18)	COOPER and RICHARDS (1966c) – Vema Seamount
Indian Ocean basin	
Island volcanics (Fig. 18)	COOPER and RICHARDS (1966b) – Reunion Island
Pacific Ocean basin	
Abyssal basalts (Fig. 18–20)	TATSUMOTO (1966b)[a]
Hawaiian volcanics (Fig. 18 – 20)	TATSUMOTO (1966a)[a]; COOPER and RICHARDS (1966b); PATTERSON (1964)
Islands volcanics, East Pacific Rise (Fig. 18)	PATTERSON and DUFFIELD (1963) – Easter Island; TATSUMOTO (1966b)[a] – Easter and Guadalupe Islands; PATTERSON and TATSUMOTO (1964) – volcanic rich sediment
Island arcs, Japan (Fig. 19)	TATSUMOTO (1966a, 1969)[a]; MASUDA (1964); COOPER and RICHARDS (1966b); SOMAYAJULU et al. (1966)[a]; KURASAWA (1968)[a]; HEDGE and KNIGHT(1969)[a]; TATSUMOTO and KNIGHT(1969)[a]
Island arcs, other areas (Fig. 19)	TATSUMOTO (1966a)[a] – Iwo Jima; COOPER and RICHARDS (1969b) – New Zealand
North America	
On or west of quartz diorite line of Moore (Fig. 20)	BANKS and SILVER (1964) – California; DOE et al.(1966)[a] – California; DOE (1968a)[a] – California; DOE (1967)[a] – California, Oregon and Washington; TATSUMOTO and SNAVELY (1969)[a] – Oregon and Washington
East of quartz diorite line of Moore but west of Idaho and Utah (Fig. 20)	This paper – Alaska and Nevada; DOE (1967)[a] – California, Oregon, and Washington; DOE (1968a)[a] – Nevada; REYNOLDS (1967) – British Columbia
Other areas (Fig. 20)	MURTHY and PATTERSON (1961)[a] – Montana; PATTERSON et al. (1955) – Nevada; STERN et al. (1965) – Utah; STACEY et al. (1969b) – Utah; DOE (1967)[a] – Montana, Idaho, Wyoming, Colorado, New Mexico, Texas and South Dakota; DOE et al.(1968)[a] – Montana; DOE et al. (1969a, b)[a] – Colorado; PETERMAN et al.(1970)[a] – Wyoming

Table 17 (continued)

Region	Sources of isotopic and concentration data
Europe	
Scotland	HAMILTON (1966); MOORBATH and WELKE (1968b)
Italy	OVERSBY and GAST (1968)[a]
Other areas of continental affinity	
East Greenland	HAMILTON (1966)
Rockall Bank, North Atlantic	MOORBATH and WELKE (1968a)
Australia and South Africa pipe rocks	LOVERING and TATSUMOTO (1968)[a]– kimberlite from Kimberley pipe, South Africa, and basaltic nephelinite from Delegate pipe, New South Wales, Australia; COOPER and GREEN (1969)–basanite of the Newer Volcanics, Western Victoria, Australia

[a] Concentration data from these papers were used in making up Fig. 15.

Table 18. *Isotopic composition of lead in abyssal basalts and in some island volcanics of the East Pacific Rise* (TATSUMOTO, 1966b)[a]

Description	$^{206}Pb/^{204}Pb$	$^{207}Pb/^{204}Pb$	$^{208}Pb/^{204}Pb$
Abyssal basalts			
Mid-Atlantic Ridge			
High alumina tholeiite (22°40'S; 13°16'W)[b]	18.47	15.54	38.01
(5°47'S; 11°25'W)[b]	17.82	15.54	37.52
(9°39'N; 40°27'W)[b]	18.82	15.68	38.65
East Pacific Rise			
High alumina tholeiite (7°47'S; 108°10'W)	18.19	15.54	37.93
(12°52'S; 110°57'W)[c]	18.24	15.53	38.03
(18°25'S: 113°20'W)	18.50	15.58	38.34
Island volcanics, East Pacific Rise			
Easter Island			
Obsidian, Mount Ourito	19.31	15.66	39.15
Andesine andesite	19.25	15.58	38.94
Alkali basalt	19.30	15.73	39.46
Tholeiite with alkalic affinities	19.28	15.67	39.16
Guadalupe Island			
Labradorite-andesine alkali basalt[d]	20.36	15.73	40.66
Labradorite alkali basalt	20.17	15.76	40.49
Labradorite olivine basalt with alkalic affinities	20.18	15.67	40.31

[a] Lead extraction by volatilization through melting. Mass spectrometric analysis uses the surface emission mode of ionization of $PbS-NH_4NO_3$ from a rhenium filament with Faraday cup collection of ions. Data are not corrected for fractionation effects in the ionization process.

[b] Average of two analyses. [c] Average of three analyses. [d] Average of two samples.

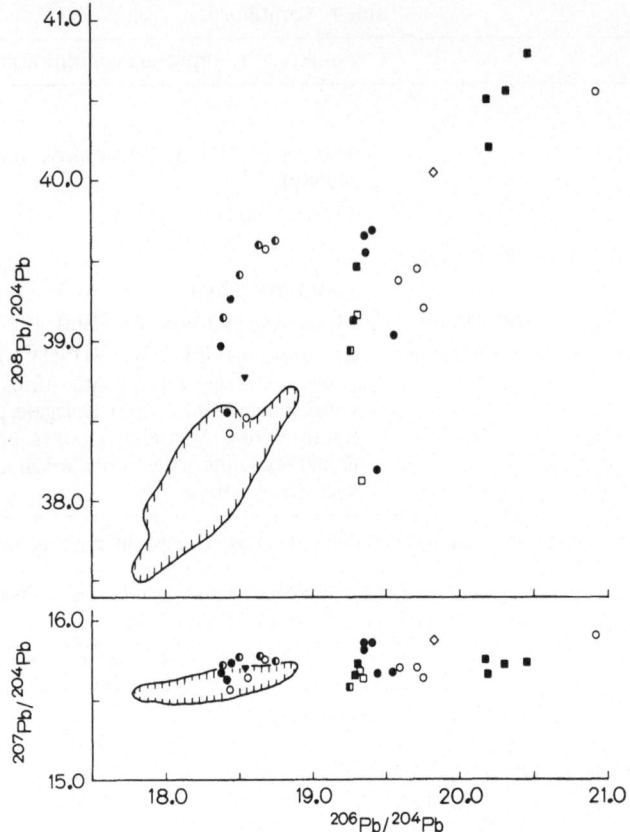

Fig. 18. $^{207}Pb/^{204}Pb$ and $^{208}Pb/^{204}Pb$ versus $^{206}Pb/^{204}Pb$ for volcanics from islands and a volcanogenic sediment from the East-Pacific Rise (■ basalt, ▣ andesite or trachyandesite, □ others), Mid-Atlantic Ridge (● basalt, ◐ andesite or trachyandesite, ○ others), Vema Seamount, S.E. Atlantic (◇ phonolite), and Reunion Islands, Indian Ocean (▼ olivine basalt). For comparison, the enclosed areas are superimposed which include all the isotopic data on abyssal basalts and volcanics of the Hawaiian Islands. References to data are given in Table 17

particularly evident on the plot of $^{208}Pb/^{204}Pb$ versus $^{206}Pb/^{204}Pb$ where the only overlap is with Iceland Group I volcanics of WELKE et al. (1968). The source of some of the island volcanics has formed relatively recently (GAST et al., 1964; TATSUMOTO, 1966b), and differentiation of U from Pb and Th from Pb has occured. Unlike the abyssal basalts of the Mid-Atlantic Ridge, there do not appear to have been major changes in $(^{238}U/^{204}Pb)$ of the source about 1,000 m.y. ago although the data of WELKE et al. (1968), when plotted on a concordia diagram, suggest some old perturbation. The slopes of their Type I and Type II volcanics diverge at the old end of the discordia lines which requires at least a three-stage history but with only minor separation of U from Pb in the second stage. GAST (1969) has developed other arguments that show changes in $(^{238}U/^{204}Pb)$ have occured over broad time spans in the sources of Island volcanics.

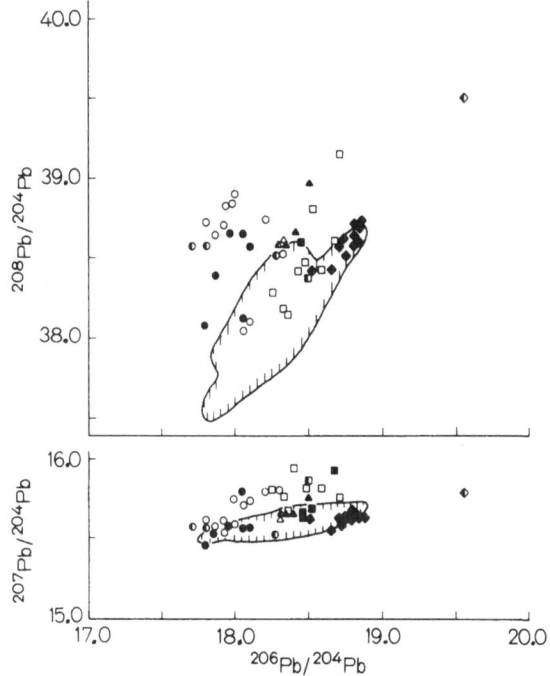

Fig. 19. $^{207}Pb/^{204}Pb$ and $^{208}Pb/^{204}Pb$ versus $^{206}Pb/^{204}Pb$ for volcanics from island arc regions: Japan (pigeonitic and hypersthenic series: ■ tholeiite, ▣ andesite, ☐ others; high alumina basalt series: ▲ high-alumina basalt, ▲ andesite, △ others; alkali basalt series: ● alkali basalt, ◐ andesite or trachy-andesite, ○ others), ◆ andesite and rhyolites of New Zealand and Iwo-Jima (◈ trachyandesite). Enclosed areas superimposed for comparison include all the isotope data on abyssal basalts and volcanics of the Hawaiian Islands. References to data are given in Table 17

Island arc data (Fig. 19; see compilation of data in Appendix D, part B) appear to be rather similar to data on abyssal basalts and open ocean features such as Hawaii although values of $^{208}Pb/^{204}Pb$ for Japanese volcanics tend to be greater than those for abyssal basalts and Hawaiian volcanics. Values of $^{207}Pb/^{204}Pb$ tend to be greater for island arc volcanics also (TATSUMOTO, 1966a; COOPER and RICHARDS, 1969b) which has been attributed to contamination of the magma lead with lead from oceanic crust, particularly that in pelagic sediments, dragged down along the Benioff zone (ARMSTRONG, 1968; TATSUMOTO,

1969). COOPER and RICHARDS (1969b) prefer some of the andesites of New Zealand to be melted geosynclinal sediments. If the lead isotope ratios for Japanese and New Zealand volcanics are representative of island arcs, Iwo Jima would fit in better with oceanic ridge and rise environments.

Continental Igneous Rocks. Various kinds of igneous rocks from the continents (Fig. 20) range widely in lead isotope ratios. Data for California, Oregon, Washington, and Nevada rocks appear similar to those for oceanic ridges and rises and are compatible with the recent conclusion, made on the basis of other evidence, that the East Pacific Rise extends under this region (YEATS, 1968). Selected data from Oregon and Washington and new data on Nevada and Alaska are given in

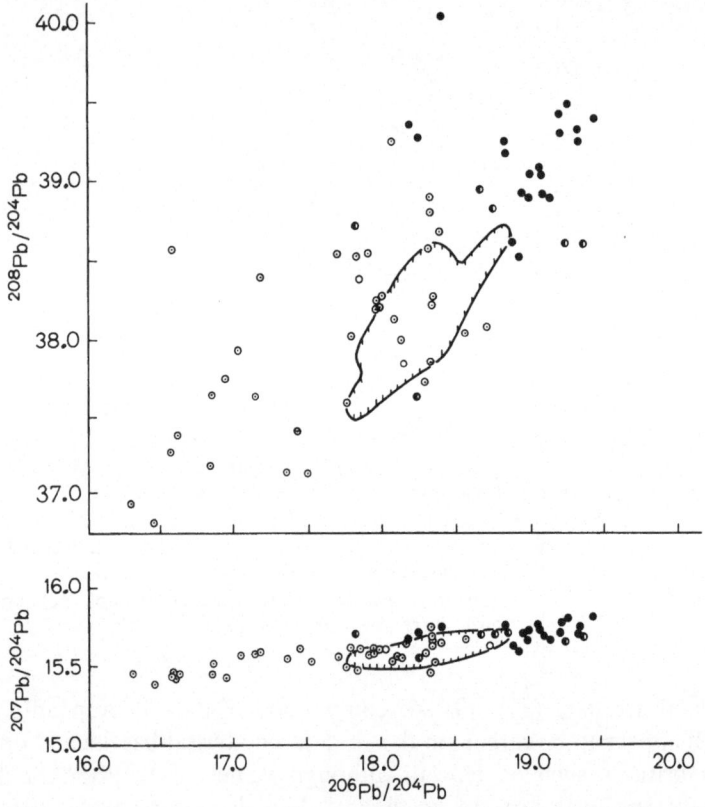

Fig. 20. $^{207}Pb/^{204}Pb$ and $^{208}Pb/^{204}Pb$ versus $^{206}Pb/^{204}Pb$ for Cenozoic and Mesozoic igneous rocks of western North America: ● California, Oregon, Washington, British Columbia, Alaska, and Nevada; ◐ Utah; ○ New Mexico, Colorado, Wyoming, Idaho Montana, South Dakota, and Texas. Enclosed areas superimposed for comparison include all the isotope data on abyssal basalts and volcanics of the Hawaiian Islands. References to data are given in Table 17

Table 19. *Selected data on Cenozoic igneous rocks from Washington and Oregon and unpublished data on feldspars from igneous rocks from Alaska and Nevada*

Sample locality	Rock type	$^{206}Pb/^{204}Pb$	$^{207}Pb/^{204}Pb$	$^{208}Pb/^{204}Pb$	Ref.
Washington					
Mount Rainier	Obsidian	19.10	15.68	39.03	1
Bodie Mountain	Obsidian	19.00	15.67	38.84	1
Pomona	Tholeiitic basalt[a]	18.42	15.76	40.03	2
Oregon					
John Day	Obsidian	19.22	15.71	39.42	1
Northwest coast	Tholeiitic basalt[a]	18.88	15.76	39.25	2
Marys Peak	Aplite dike	19.11	15.73	39.30	2
gabbroic sill	Pegmatite	19.25	15.75	39.26	2
	Granophytic diorite	19.22	15.74	39.30	2
	Basalt, chilled border	19.31	15.75	39.34	2
Widow Creek	Alkalic basalt[a]	19.34	15.70	39.32	2
Coffin Butte	Tholeiitic basalt	18.98	15.61	38.74	2
Alaska, Seward Peninsula					
Hume Creek (64°54′N; 166°5′W)	Granite	18.98	15.72	38.93	3
Brooks "Range" (65°31.5′N; 167°10′W)	Granite	19.09	15.72	39.13	3
Nevada					
Busted Butte (36°47′N; 116°25′W)	Rhyolite vitrophyre	18.18	15.69	39.42	3
	Quartz latite pumice	18.23	15.66	39.29	3
Pahute Mesa (37°16′N; 116°24′W)	Rhyolite vitrophyre	18.32	15.68	39.24	3
	Quartz latite derivative	18.41	15.72	39.30	3

References:
(1) DOE (1967); after dissolution by HF-$HClO_4$, mass spectrometric analysis was made from PbS-NH_4NO_3 by the surface emission mode of ionization from rhenium filament with Faraday cup collection of ions. Data are normalized to agree with Ta-1 filament material to bring ratios within stated analytical uncertainties of Tilton's 1:1 lead isotope gravimetric standard.
(2) TATSUMOTO and SNAVELY (1969); lead extraction by volatilization below the melting point with other analytical factors as given in Table 18.
(3) This paper; analytical factors as given in reference 1 above. K-feldspar separates were provided by C. L. SAINSBURY for Alaska and PETER LIPMAN for Nevada; both donators are from the U.S. Geological Survey.

    [a] Average of two samples.

Table 19. Igneous rocks formed in Precambrian crystalline terranes tend to be unradiogenic in their initial lead isotope ratios relative to oceanic ridges and rises which may be due to ancient metamorphism of the source rocks which reduced the U/Pb (HAMILTON, 1966). In North America, the unradiogenic leads are concentrated in the Rocky Mountain region and farther east. Present data on igneous rocks indicate

Table 20. *Selected lead isotope data on basalts and silicic igneous rocks of Cenozoic age in the Rocky Mountain region, United States*[a]
(Abbreviations: prim.: primitive; cont.: contaminated)

Sample locality	Sample type	Isotopic ratios			Ref.
		$^{206}Pb/^{204}Pb$	$^{207}Pb/^{204}Pb$	$^{208}Pb/^{204}Pb$	
Snake River Plain, Idaho					
American Falls	Obsidian	18.36	15.76	38.80	1
Shoshone Falls	Basalt (prim.)	18.12	15.45	38.08	4
		18.35	15.62	38.88	5
Craters of the Moon	Basalt and andesite (cont.)[b]	17.89	15.64	38.75	2,5
San Juan Mountains, Colorado					
Creede area	Silicic volcanic	18.74	15.63	38.09	1
Wagon Wheel Gap area	Silicic volcanic	18.36	15.58	37.68	1
La Jara Reservoir area	Hinsdale Formation				
	Basalt (prim.)	18.33	15.55	37.67	3
	Basalts (cont.)[b]	17.94	15.54	37.48	3
	Basalt (cont.)	18.38	15.62	38.26	1
Jemez Mountains, New Mexico					
Los Posos	Obsidian	18.01	15.57	37.99	1
Arroyo Hondo	Obsidian	18.20	15.53	38.00	1
Borrego Mesa	Basalt (prim.)[b]	18.34	15.46	37.86	1
Cerro Pelon	Basalt (cont.)[b]	17.78	15.50	37.60	1
Rio Grande Gorge, New Mexico					
Taos area	Serviletta Formation of Montgomery (1953)				
	Basalt (prim.)	18.09	15.51	37.55	3
	Basalt (cont.)[b]	17.37	15.50	37.21	3

References: (1) DOE (1967); (2) DOE (1968a); (3) DOE et al. (1969b); (4) PATTERSON et al. (1955); (5) TILTON, in DOE (1968a).

[a] All silicic samples decomposed by HF–HClO₄ dissolution. Basalts were treated by volatilization extraction with three checked by HF–HClO₄ dissolution. All data except those of Ref. 4 are normalized to agree with Ta-1 filament material with is within the error limits indicated by the Tilton 1:1 lead isotope gravimetric standard.

[b] Average of two samples.

that Utah is an isotopic transition zone between Nevada and the Rocky Mountain region (Fig. 20), rather than a part of the Rocky Mountain region grouping as previously stated (DOE, 1968a). Basalts from the Rocky Mountain region are rather similar in lead isotopic composition to silicic igneous rocks, even if obviously contaminated basalts are eliminated (Table 20). These data suggest that both rock types come from a similar age source with similar values of U/Pb and Th/Pb. This source may be the lower crust or the low-velocity channel of the upper mantle, and the silicic rocks may come from pods of more silicic material in the eclogite or granulite facies (DOE, 1968a). Contrary to the isotopic relations in the Rocky Mountain region, in Scotland the mafic igneous rocks statistically appear to be more radiogenic than the silicic igneous rocks of Cenozoic age, MOORBATH and WELKE (1968b). The silicic igneous rocks, which are completely overlapped in their lead isotope ratios by those from whole-rock samples from the nearby Lewisian basement (MOORBATH et al., 1969), might be interpreted as being derived from the lower crust with the mafic igneous rocks coming from the mantle and contaminated to varying degrees with crustal material. MOORBATH and WELKE (1969), however, prefer a mantle source with crustal interaction for all rock types but with greater crustal contribution in the silicic rocks. The true significance of the difference in isotopic behavior of mafic and silicic igneous rocks between the Rocky Mountain region and Scotland remains a problem.

Ultrabasic Rocks. Although a number of determinations of lead isotope ratios have been reported for ultrabasic rocks, only those for the kimberlite from South Africa (LOVERING and TATSUMOTO, 1968) and peridotitic portions of a layered mafic intrusion in Scotland (MOORBATH and WELKE, 1968b) can be considered reliable whole-rock values. In view of the great importance of data on these kinds of rocks and the difficult analytical problems, only those data considered reliable whole-rock or mineral values are included in Table 21.

Lead isotope data are still valuable even when the lead is only partially extracted and its lead isotopic composition is not equal to that of the whole-rock. A Pb-Pb isochron age may still be calculated and a legitimate comparison may be made of lead from possibly isotopically heterogeneous inclusions with that from isotopically uniform host rocks of Mesozoic or Cenozoic age. COOPER and GREEN (1969) have made a well executed study of this kind on lherzolite inclusions and host alkali basalt from western Victoria, Australia. Partly volatilized lead appears to always be more radiogenic than the whole-rock value on old rocks. As the volatilized leads analyzed by COOPER and GREEN from lherzolite are less radiogenic than those of the isotopically uniform host basalt, the data require separate immediate source regions for the leads in the

Table 21. *Isotopic composition of lead in ultrabasic and related rocks*

Sample type		$^{206}Pb/^{204}Pb$	$^{207}Pb/^{204}Pb$	$^{208}Pb/^{204}Pb$	Ref.
Kimberley pipe, South Africa					
Kimberlite	(initial ratios)[a]	19.51	15.82	39.32	1
	(present-day ratios)	19.77	15.83	39.66	1
Cuillins layered intrusion, Scotland					
Gabbro, Zone I	(leucocratic band)	17.84	15.69	37.97	2
	(melanocratic band)	17.77	15.67	37.84	2
Gabbro, Zone IV		17.60	15.55	38.00	2
Peridotite, main body		18.13	15.66	38.44	2
Sligachan area, Scotland					
Peridotite dike in basalt		17.86	15.58	38.02	2

References: (1) LOVERING and TATSUMOTO (1968); (2) MOORBATH and WELKE (1968b).

[a] Corrected for 174×10^6 yrs. of decay of uranium and thorium.

two rock types. This conclusion may be extended to the rocks themselves if the alkali basalt has not been contaminated with lead that has been well mixed through the magma subsequent to inclusion of the lherzolite. Although only a preliminary report is available, MANTON and TATSUMOTO (1969) find somewhat similar results between eclogite inclusions and host kimberlite from the Roberts Victor Mines in South Africa. Both the lherzolites from Australia and the eclogites from South Africa are calculated to come from sources 2,000–2,500-m.y.-old. GRAESER and HUNZIKER (1968) found that the values of $^{207}Pb/^{204}Pb$ in ultrabasic rocks from the Ivrea Zone in northern Italy are less than those in galenas from the nearby Alpine region. Different sources for the ores and the ultrabasic rocks are suggested by this difference.

Mediterranean Volcanics. Few lead isotope data are available on Mediterranean volcanics (see compilation in Appendix D, part C). If only the lead isotope data are looked at, these volcanics would fall in a grouping with oceanic ridges and rises and the coastal area of Western North America. Volcanologists specializing in the region have tended to classify these volcanics as a distinct class. As the isotopic data are fragmentary, the distinct classification is adopted.

8. Paleozoic and Precambrian Igneous Rocks

Igneous rocks older than Mesozoic are somewhat difficult to work on as the corrections for in situ radioactive decay of U and Th are large and the values of U/Pb and Th/Pb often have undergone a recent

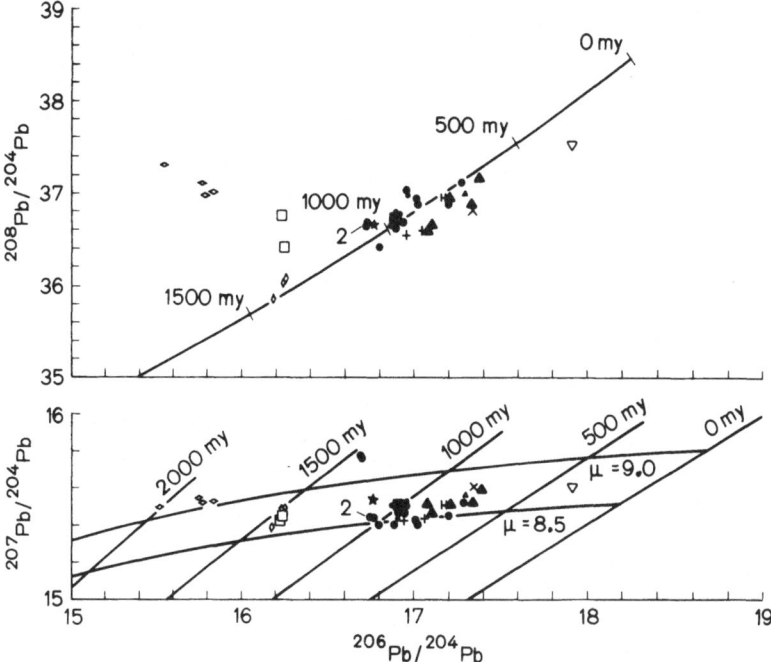

Fig. 21. Lead isotope relationships for feldspar leads from 1,000-m.y.-old igneous rocks. Curved lines on the plot of $^{207}Pb/^{204}Pb$ versus $^{206}Pb/^{204}Pb$ are the single stage growth curves of $^{238}U/^{204}Pb$ for values of 9.0 and 8.5, and the heavy sloping lines are the isochrons for the labeled ages. The normal growth curve involving $^{208}Pb/^{204}Pb$ and $^{206}Pb/^{204}Pb$ is given in that plot with positions of model leads indicated for various ages. The plot is from ZARTMAN and WASSERBURG (1969) and includes their data from: ● Texas suite, ▲ Shenandoah National Park, Va., ▲ Western Adirondack Mountains, N.Y., ▽ Eastern Adirondack Mountains, N.Y., + Westport, Ontario, ★ Leg Pond, Newfoundland, □ ■ Port Cartier-Mt. Reed, Quebec, H Herefoss, Norway, · Pikes Peak, Colorado, × Duluth, Minnesota, ◇ Mellen, Wisconsin, ◇ Gold Butte, Nevada; as well as the following data: ● Llano, Texas (ZARTMAN, 1965b), ▲ Shenandoah National Park, Virginia (DOE et al., 1965), ▲ Balmat, New York (DOE, 1962b), ·Pikes Peak, Colorado (DOE, 1967)

disturbance. The more successful approach is to analyze the lead in a phase with low values of U/Pb and Th/Pb such as K-feldspars. Such data for rocks about 1,000-m.y.-old are given on Fig. 21 (ZARTMAN and WASSERBURG, 1969). Selected data are given in Table 22. The possible addition of lead from metamorphic effects has obscured the cause of the fact that in North America, Precambrian and Paleozoic igneous rocks have acquired a progressively greater excess radiogenic increment with decrease in time (DOE et al., 1965). This is true even of most granitic rocks found in terranes that are much older than the granitic rock. North American leads unradiogenic for their age as in Mesozoic and

Table 22. Selected feldspar lead data on Paleozoic and Precambrian rocks (corrections for decay of U and Th are less than 2 percent)

Sample locality	$^{206}Pb/^{204}Pb$	$^{207}Pb/^{204}Pb$	$^{208}Pb/^{204}Pb$	Ages (m.y.)			Ref.
				Isochron model	$^{208}Pb/^{204}Pb$ model	Accepted	
Some K-feldspars in which model lead ages are in good agreement with accepted ages							
Pikes Peak batholith, Colorado	16.91	15.49	36.71	980	980	1050	1
Commercial pegmatite, Wyoming	14.62	15.27	34.24	2530	2410	2450	2
Microcline granite, Kola Peninsula, Baltic Shield	13.65	14.59	33.50	2750	2660	2800	3
Some K-feldspars in which model lead ages are older than accepted ages							
Pegmatite, Pitkyaranta, Baltic Shield	14.53	15.09	34.43	2450	2220	1900	3
Rapakivi granite, Nevada[a]	16.21	15.44	35.99	1450	1350	1100	4
Granite, Wisconsin[b]	15.73	15.51	37.08	1870	770	1100	4
Some K-feldspars in which model lead ages are younger than accepted ages							
Pegmatite in Baltimore Gneiss, Maryland	18.60 / 18.50[c]	15.75 / 15.62[c]	38.65 / 38.17[c]	−80	20	450	5
Charnockite, Sabarov, Ukraine	15.97	15.27	36.05	1450	1260	2200	3
Feldspars with at least one ratio reasonably unaffected by metamorphism							
Granite gneiss, Pitkyaranta, Baltic Shield	14.35	15.13	33.51	2600	2320	2600(1800)[d]	3
Baltimore gneiss, Maryland	17.42 / 18.10	15.52 / 15.53	40.01 / 37.09	610 / 70	−890 / 770	1100(300—450)[d]	5

References: (1) Doe (1967); (2) Catanzaro and Gast (1960); (3) Tugarinov et al. (1963a); (4) Zartman and Wasserburg (1969); (5) Doe et al. (1965).

[a] Average of 3 samples. — [b] Average of 4 samples. — [c] Muscovite.
[d] Age in ( ) is the age of metamorphism. The other age is the accepted age of the rock.

Cenozoic igneous rocks have been much more difficult to find. The situation does not appear to be much different in Europe or Asia where most feldspar leads give model lead ages equal to or younger than their known age such as the data of KOUVO (1958) in Finland and most of the data of TUGARINOV et al. (1963a) on the Baltic, Ukrainian, and Aldan Shields. Existence of older equivalents of the Rocky Mountain type seems likely. The ones that have been found are small, shallow intrusions (ZARTMAN and WASSERBURG, 1969) and in a pegmatite from the Baltic Shield at Pitkyaranta in Southwest Karelia (TUGARINOV, et al., 1963a). Though many old granitic whole-rock samples of the Aldan Shield are unradiogenic in their present-day ratios, no data yet demonstrate that any are very unradiogenic in their initial ratios. TUGARINOV et al. (1963a), for example, give data on two feldspars from 1,900-m.y.-old rocks in the Stanovoy Range of the Aldan Shield in which both are in reasonable agreement with the accepted age. Apparently, most older granitic rocks form in a terrane somewhat similar to that of the present west Coastal States of the U.S. or oceanic ridges or rises.

9. Ore Genesis

General. Emphasized in this section are sulfide ore bodies and natural solutions which are normally not radioactive. Uranium deposits are not included because they generally do not contain common lead and few data are yet available on other sorts of deposits, such as iron ore and gold. The bearing of lead isotopes on genesis of sulfide ore bodies is considered.

The genetic categories chosen for the isotopic considerations are: magmatogenic, indirect magmatogenic (which might be called lateral secretion with magmatogenic influence), regional metamorphic origin, lateral secretion (without known magmatic influence), and syngenetic-diagenetic origin. Particular attention is given to stratiform deposits because lead isotope data clearly indicate more than one process for their generation.

Magmatogenic Ores. Ores in which the isotopic composition of ore lead is the same as that in the presumably related igneous source at the time of crystallization are good candidates for a magmatogenic origin. Data for this "fingerprint" are available in support of this mechanism. These ores are usually isotopically uniform with $^{206}Pb/^{204}Pb < 19.5$ and $^{208}Pb/^{204}Pb < 39$. Examples (Fig. 22) are the ores at Butte, Montana (DOE et al., 1968), Bingham Canyon, Utah (STACEY et al., 1968), and the Nelson batholith, British Columbia (REYNOLDS, 1967). It must be

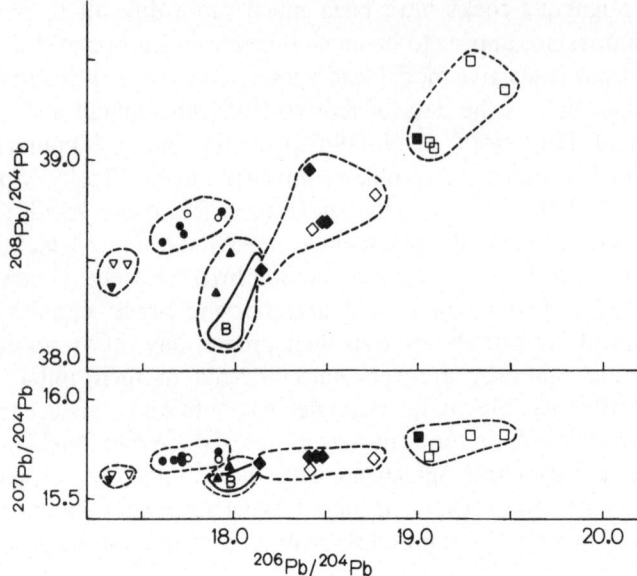

Fig. 22. $^{207}Pb/^{204}Pb$ and $^{208}Pb/^{204}Pb$ versus $^{206}Pb/^{204}Pb$ for some sulfide ores with isotopic support for a direct magmatogenic origin of the ores through comparison with the initial lead of possibly related igneous rocks as estimated by feldspar leads: galenas from the U.S. and Lark mines (●) with the Tertiary Last Chance stock at Bingham Canyon (○) and galena from the Star and San Francisco mining districts (◆) with Tertiary granitic rocks of the Milford region (◇) (STACEY et al., 1968); galena ore composites (▲) with the Butte Quartz Mon-zonite and related igneous rocks (12 samples in area labeled B) (DOE et al., 1968); a galena veinlet in the Rader Creek pluton (▼) with the Donald pluton (▽) of the Cretaceous Boulder batholith, Montana (DOE et al., 1968); galena from the Scranton mine (■) in the Jurassic Nelson batholith (□), British Columbia (REYNOLDS, 1967)

remembered that in cases of extreme wallrock alteration, such as at Butte, that the lead may have been leached from the igneous rock during alteration rather than be a product of magmatic differentiation.

Indirect Magmatogenic Ores. Many ores thought to be derived in the igneous process have lead isotopic compositions different from the likely igneous sources of the region or of even larger areas. The lead (Fig. 23) is usually more radiogenic than the igneous rocks such as the ores of the Milford region, Utah (STACEY et al., 1968), although such deposits around the Nelson batholith are less radiogenic (REYNOLDS, 1967). The isotopic data demand that a large component of the ore lead be derived from sources outside the magma and be lead probably generated by radioactive decay of U and Th from rocks of a much older terrane. The facts that the igneous rock has usually the least radiogenic lead and that the ores form a continuum of more radiogenic values

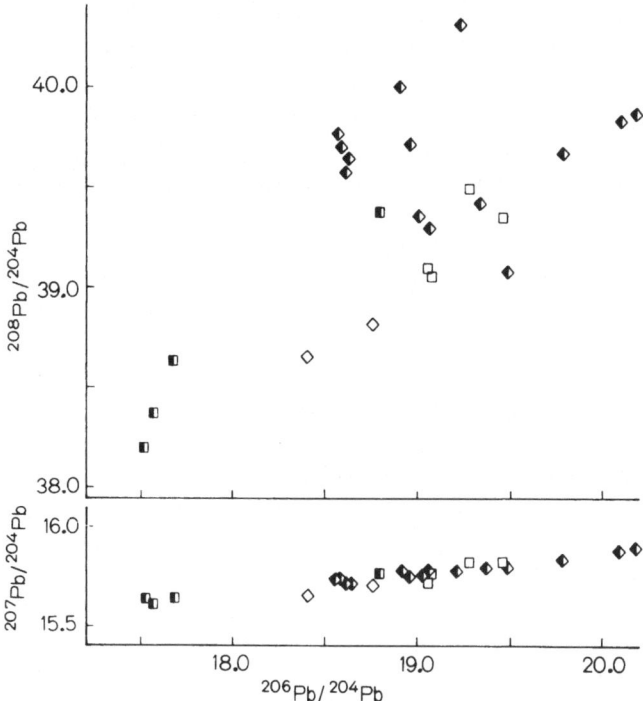

Fig. 23. $^{207}Pb/^{204}Pb$ and $^{208}Pb/^{204}Pb$ versus $^{206}Pb/^{204}Pb$ for ores in which isotopic differences between galena ores and igneous rocks (as estimated by their feldspars) give evidence that some or all the lead in the ores must have been derived from outside the magma chamber (indirect magmatic): Tertiary granitic rocks of the Milford region (◇), Utah, with galenas from mines and prospects in the Star, Wah Wah, Lincoln, and Salina Creek mining districts (◆) (STACEY et al., (1968); granitic rocks of the Jurassic Nelson batholith (□), British Columbia, with galenas from the nearby Blue Star, Lakeshore, Victor, and Bluebell mines (▥) (REYNOLDS, 1967)

suggest that the igneous rock supplied material as well as heat during ore-formation and that there was a magmatic component of lead in the ore. This process is akin to lateral secretion in the resultant isotopic characteristics but may have even greater isotopic variability. These data allow calculation of Pb-Pb ages to determine the age of the source of the lead. The agreement between calculated age and age of postulated source is often good (KANASEWICH, 1962; HEYL et al., 1966; STACEY et al., 1968).

Deposits of Metamorphic Origin in Much Older Rocks. Lead isotope data clearly support a metamorphic origin for some deposits of this kind although well-documented studies are few for this little-emphasized category. In order to have a good isotopic solution, the metamorphism must have acted upon much older rocks to allow for isotopic variability. The characteristics are, again, of the lateral secretion type, but even greater variability in isotope ratios may be encountered. The ratios may range from those expected for the time of formation of the source rocks to those much more radiogenic than expected for the age of metamorphism. A Pb-Pb age may be calculated to determine the age of the source terrane if the metamorphic age is known or visa versa. The results may be confirmed by other dating methods. The Thackaringa deposits (RUSSEL et al., 1961) in Australia are the best example of this type.

Deposits Possibly of Metamorphic Origin in not Much Older Sediments. Many very large stratiform sulfide deposits may have formed through metamorphism of sediments composed of volcanic or calcareous materials not much greater in age than the metamorphism. The leads often have isotopic compositions approximating the *normal growth* or *single stage* conditions but with model lead ages in better agreement with the age of metamorphism rather than of the age of sedimentation. KANASEWICH (1968a) has referred to this condition as "short-period anomalous lead". Bathurst, New Brunswick, Canada (Table 16), is one possible example as the ore is in Middle Ordovician rocks (approximately 450-m.y.-old) that were metamorphosed at about 375 m.y. ago. The isochron model lead age is 270 m.y., which is within uncertainties of the parameters of the equations of the growth curve and of the age of metamorphism. If the age of the earth is chosen to be only 20 m.y. less than the accepted value, the isochron model lead age is increased by 90 m.y. at this geologic time. Deposits near Lake Manitouwadge, Ontario, Canada, have been suggested to be of this derivation (OSTIC et al., 1967), as have been the deposits near Sullivan, British Columbia, Canada (SINCLAIR, 1966). Balmat, New York, has deposits that might be of this derivation (DOE, 1962b); however, recent work by REYNOLDS and RUSSELL (1968) has shown a systematic shift in ^{207}Pb/^{204}Pb between the ores on the one hand and some of the granitic rocks and marble on the other. All these deposits also fit the category developed of *conformable* ore bodies (STANTON and RUSSELL, 1959) of a possible syngenetic relationship of ore with island arc or continental margin tectonics. The alternative explanation expressed here must still be considered.

Lateral Secretion Through Much Older Rocks. Many important stratiform ores, such as those of the Mississippi Valley type, do not have any close association with either volcanism or metamorphism. (Some lead isotope data are summarized in Table 23, Fig. 24.) These ores are characterized by isotopic heterogeneity, as first shown by NIER (1938) and NIER et al. (1941). The lead isotope data nicely fit a lateral secretion

Table 23. *Stratiform ores of Mississippi Valley-type related to leaching of much older rocks. (Ores are in Paleozoic limestones.)*

District of the United States	Kind of ratio (no. of samples)	$^{206}Pb/^{204}Pb$	$^{207}Pb/^{204}Pb$	$^{208}Pb/^{204}Pb$	Ref.
1. Illinois-Kentucky fluorite	Minimum	19.91	15.80	39.92	2
	Maximum	21.01	15.98	40.64	
	Average (8)	20.36	15.90	40.27	
2. Southeast Missouri galena	Minimum	19.96	15.65	39.10	1
	Maximum	21.46	16.19	41.02	
	Average (131)	20.81	15.97	40.08	
3. Tri-State sphalerite	Minimum	21.31	16.00	41.07	3
	Maximum	22.66	16.28	41.92	
	Average (14)	22.11	16.15	41.46	
4. Wisconsin-Illinois-Iowa sphalerite-galena	Minimum	20.83	15.96	40.45	2
	Maximum	24.44	16.33	43.95	
	Average (19)	22.54	16.15	42.31	

References: (1) BROWN (1967); (2) HEYL et al. (1966); (3) RUSSELL and FARQUHAR (1960).

mechanism for these deposits, perhaps with the source of the lead being the basement rocks, as suggested by KANASEWICH (1962) and HEYL et al. (1966), because of the similarity of the Pb-Pb age to that accepted for the basement. These ores characteristically are highly radiogenic with isotope ratios greater than those initially present in any igneous rock of any kind, anywhere in the world. KANASEWICH (1962) has concluded that tectonic uplift caused the ore fluids to move.

Ores Possibly Related to Sediments of Nearly Equivalent Age. Stratiform deposits of another important type that is isotopically distinct from the Mississippi Valley type have uniform lead isotopic compositions which are in reasonably good agreement with the known age of the wallrocks. One excellent example is the Kupferschiefer Marl-Slate (Table 16). Many investigators favor a syngenetic origin for these deposits although WEDEPOHL (1964) has shown that the syngenetic fluids which deposited the Kupferschiefer Marl-Slate must have been unusually rich in heavy metal content relative to ocean water.

Some other deposits of this type are also included in the category of a genetic relation (perhaps syngenetic from volcanic emanations) with the formation of island arcs and continental margins as proposed by R. L. STANTON (in STANTON and RUSSELL, 1959). According to current tectonic thought, such regions involve downwarping of oceanic crust into the mantle. Pelagic sediments are the only rock type with the proper isotopic composition at present to qualify for forming such deposits today (BROWN, 1965). ARMSTRONG (1968) has proposed this

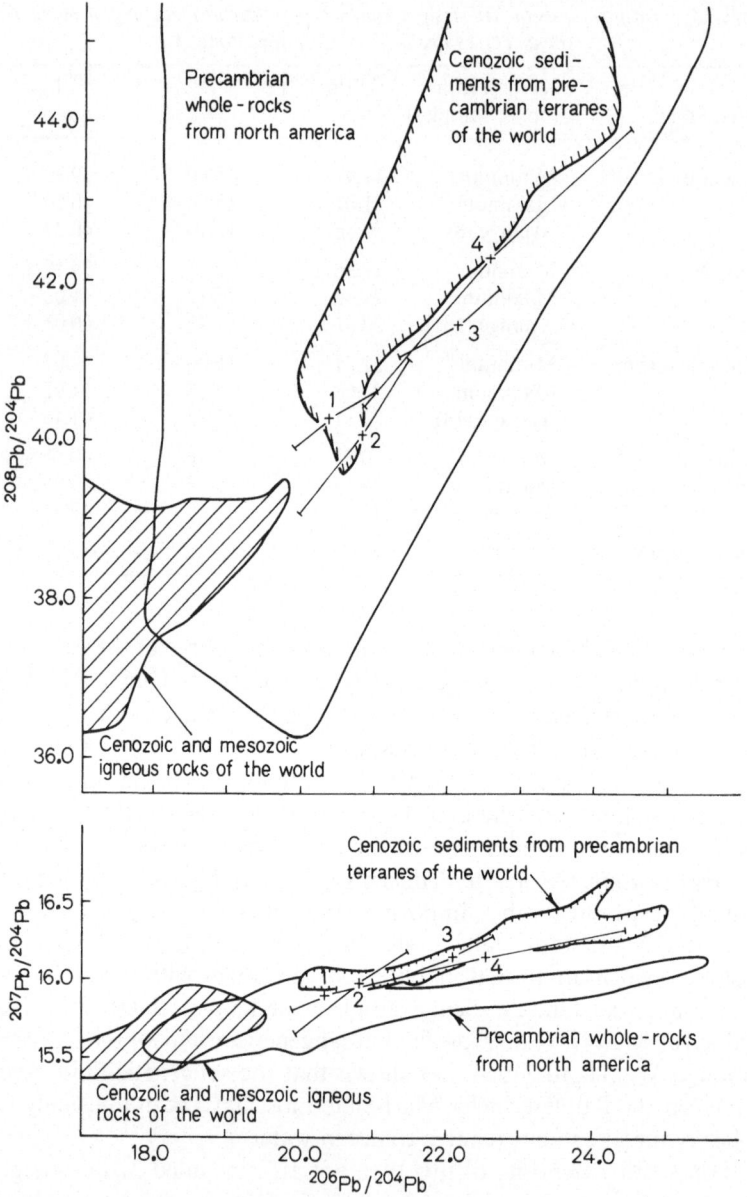

Fig. 24. ^{207}Pb/^{204}Pb and ^{208}Pb/^{204}Pb versus ^{206}Pb/^{204}Pb for extremes and means of four Mississippi Valley districts: **1** Illinois-Kentucky fluorspar district (HEYL *et al.*, 1966); **2** Southeast Missouri lead belt (BROWN, 1967); **3** Tri-State zinc district (RUSSELL and FARQUHAR, 1960); **4** Wisconsin-Illinois-Iowa lead-zinc district (HEYL *et al.*, 1966). For comparison, enclosed areas include the kinds of data as labeled

downwarping mechanism along the Benioff zone as the mechanism of getting the lead out of the pelagic sediments (mixed with some other lead derived from mafic rocks) onto the continents. In contrast, KANASEWICH (1968b) has proposed that many of these ores are related to zones of pull-apart such as the Red Sea and California's Salton Basin. This condition differs from the island-arc hypothesis in that such zones probably constitute upwelling of new mantle material. The attraction of this hypothesis is that the heavy metal-rich, chloride-rich brines have been found in these pull-apart zones and may be the ore fluid for many of these deposits. OSTIC et al. (1967) have favored a mantle source for the lead in many of these deposits.

Vein Deposits of Uncertain Origin. Some vein deposits with uncertain origin are extremely heterogeneous in lead isotopic compositions and can be highly radiogenic; among these are deposits in the Thunder Bay region (FARQUHAR and RUSSELL, 1957). Inasmuch as the source rocks of the mineralization near Thunder Bay undoubtedly belong to the Superior Province and are about 2,800-m.y.-old, the Pb-Pb age calculation suggests that the mineralization was quite recent. The mechanism involved in forming these veins is not certain except that leaching of lead from much older rock must be involved.

Other vein deposits have lead isotopic compositions that closely approach a single stage generation. The origin of these deposits is problematical. In some localities, such as Aouli in North Africa, they may be related to formation of the nearby stratiform deposits at Mibladen, Morocco (CAHEN et al., 1958); however, there is not a stratiform deposit reported from the entire Churchill Province in North America with which to relate the Flin Flon vein deposit.

Brines and Ores Possibly Related to Brines. The chloride-rich brines that carry great concentrations of heavy metals are a relatively new discovery and may be the ore fluid for some of the categories mentioned above. Several vein and stratiform ores near the Red Sea could well be related to the Red Sea geothermal brines, for example (Table 24). COOPER and RICHARDS (1969a) suggest that the variation in lead isotopic composition of acid soluble lead in the deeper sediments may be due to contamination of the brine lead with lead from pelagic sediments. This kind of explanation might also account for the variation of lead isotope ratios of the stratiform deposits of Egypt where the brine lead may have become contaminated with more radiogenic wallrock lead during emplacement. The new data available so far show the lead in the brines to be somewhat radiogenic (DOE et al., 1966; DELEVAUX et al., 1967). Some deposits usually thought to be of syngenetic origin may be related to these or other exotic fluids that empty out on the ocean floor or deeps.

Table 24. *Isotopic composition of lead in brines, precipitate from brines, and vein and stratiform ores possibly related to brines*

Region	Lead in	No. of samples	Isotopic ratios[a]			Ref.
			$^{206}Pb/^{204}Pb$	$^{207}Pb/^{204}Pb$	$^{208}Pb/^{204}Pb$	
Salton Trough geothermal area, California	Chloride brine	4 (avg.)	19.13	15.67	39.07	2
Red Sea, Africa (Atlantic II deep)	Chloride brine	1	18.72	15.68	38.52	1
Red Sea, Africa	Surface precip.	1	18.68	15.75	38.82	3
	Deeper precip. from hot spots	12 (avg.)	18.92	15.77	39.08	3
Red Sea, Africa	Pelagic sediments	9 (avg.)	18.95	15.79	39.32	4
Rabigh, Saudi Arabia	Galena from barite vein	1 1	18.80 18.787	15.68 15.683	38.53 38.503	1 1
Egypt	Galena from stratiform deposits					
Bir Ranga		1	18.71	15.74	38.90	1
Um Ans		1	19.20	15.81	39.36	1
Um Gheig		1	19.28	15.77	39.17	1

References: (1) DELEVAUX et al. (1967); DOE et al. (1966); (3) COOPER and RICHARDS (1969a); (4) CHOW (1968b).

[a] All samples were analyzed by surface emission mass spectrometry except the sample that is reported to an extra figure (from Rabigh) which was analyzed by lead tetramethyl mass spectrometry. All precipitate and sediment data are on acid soluble lead only and are not whole-rock values.

Economic Considerations. Surprisingly little practical application of lead isotopes has been made by the mining industry in spite of several highly successful academic studies that demonstrate their usefulness. Use of lead isotopes in prospecting for old uranium deposits is outlined in CANNON et al. (1958). A statistical approach to evaluating ore bodies not of the Mississippi Valley type is given by CANNON et al. (1961). The use of lead isotopes as "fingerprints" to compare mineral prospects with active mines in a district is outlined in DELEVAUX et al. (1967). STACEY et al. (1968) report that the large ore bodies in a district have isotopic compositions closest to those of the igneous rock most likely to be the source. There seems to be little doubt of broad utility in the use of lead isotopes in mineral prospect evaluation.

10. Isotopic Composition of Lead in Natural Waters and in the Atmosphere

Oceans. The isotopic composition of lead has not yet been measured directly in ocean water because of the low lead content; however, the isotopic composition of HCl-soluble lead in manganese nodules (CHOW and PATTERSON, 1959) is commonly accepted to represent that of lead in the oceans (Table 25). The lead isotopic compositions of manganese nodules are also not much different from those in the pelagic sediments.

Brines. Some data are available on the direct measurement of lead isotope ratios in chloride-rich natural brines (Table 24). These brines contain unusually large heavy metal contents and are discussed in the context of an ore fluid in the previous section.

Continental Waters. Some data have recently been obtained for Central Kazakstan (Table 25) on waters in cracks and sediments of Devonian volcanics and sediments and water from the Sea of Azov and Lake Balkash (ANDREYEV *et al.*, 1967). In this difficult analitical problem, the lead was collected either (1) on activated charcoal saturated with chloroform saturated with dithizone or (2) as a coprecipitate with a tannin-gelatin complex. The transient water was found to be highly variable in lead isotope ratios, both radiogenic and non-radiogenic and the lake waters had about the same ratios as oceanic manganese nodules or were less radiogenic.

Atmosphere. Few data are available (Table 25) on the isotopic composition of lead in the atmosphere. The low lead content of air requires some concentrating mechanism before analysis is feasible, and snow has been used (TATSUMOTO and PATTERSON, 1963; CHOW, 1968a). This lead may be due to lead from gasoline widely carried as aerosols (CHOW and JOHNSTONE, 1965), as they are rather similar in isotopic composition.

Table 25. *Isotopic composition of lead in natural waters and in the atmosphere*

Locality (reference)	Material Analyzed	No. of samples	$^{206}Pb/^{204}Pb$	$^{207}Pb/^{204}Pb$	$^{208}Pb/^{204}Pb$
Arctic Ocean water (1)	Jurassic Mn nodule	1	18.78	15.84	39.33
Atlantic Ocean water (1)					
N.E. Atlantic, area F in (1)	Cenozoic Mn nodules	3	19.06	15.82	39.58
N.W. Atlantic, area G in (1)	Cenozoic Mn nodules	7	19.13	15.82	39.66

Table 25 (continued)

Locality (reference)	Material Analyzed	No. of samples	$^{206}Pb/^{204}Pb$	$^{207}Pb/^{204}Pb$	$^{208}Pb/^{204}Pb$
Pacific Ocean water					
West Pacific (1), area A in (1)	Cenozoic Mn nodules	2	18.55	15.67	39.00
West Pacific (2), area A in (1)	Cenozoic Mn nodules	4	18.51	15.62	38.80
North Pacific (1), area B in (1)	Cenozoic Mn nodules	8	18.81	15.79	39.28
East Pacific (1), area C in (1)	Cenozoic Mn nodules	10	18.85	15.78	39.29
S.E. Pacific (1), area D in (1)	Cenozoic Mn nodules	4	18.99	15.88	39.47
Gulf of California (1)	Cenozoic Mn nodules	1	19.20	15.73	38.98
Central Kazakstan waters (3)					
Sea of Azov	Water	1	18.08	15.05	36.79
Lake Balkash	Water	1	18.65	15.70	38.18
From cracks in volcanics and	Water	2	17.38	14.79	35.89
Devonian sediment rocks[a]		3	18.37	15.92	38.74
		5	19.28	16.04	38.58
		1	21.54	16.28	39.92
		2	25.53	15.91	37.58
		1	198.7	34.57	51.43
Atmosphere					
California (4)	Snow	1	18.01	15.74	38.40
Greenland (5)	Snow	1	18.41	15.64	38.66

References: (1) CHOW and PATTERSON (1962a); (2) CHOW and TATSUMOTO (1964); (3) ANDREYEV et al. (1967); (4) TATSUMOTO and PATTERSON (1963); (5) CHOW (1968a)

[a] Averages of samples included in the following ranges of $^{206}Pb/^{204}Pb$: 17—18, 18—19, 19—20, 20—25, 25—26, greater than 26.

IV. Radioactive Lead Isotopes

There are four radioactive isotopes of lead: ^{210}Pb and ^{214}Pb in the ^{238}U decay chain, ^{211}Pb in the ^{235}U decay chain, and ^{212}Pb in the ^{232}Th decay chain. All except ^{210}Pb have half-lives of less than 12 hours. Such short half-lives are of limited usefulness in geology; therefore, ^{210}Pb with a half-life of 22.0 ± 0.5 yr. (HOUTERMANS et al., 1964; RAMTHUN, 1964; GUTEN et al., 1967) has received the major interest. The analytical techniques for ^{210}Pb are given in GOLDBERG (1963).

The ^{210}Pb has been found useful in studies of volcanic rocks where it is often found in disequilibrium with ^{238}U in recent volcanics (SOMAYAJULU et al., 1966; OVERSBY and GAST, 1968b) and in fumarolic matter (SOMAYAJULU et al., 1966; HOUTERMANS et al., 1964). These studies show promise of dating recent volcanism within the last 100 yrs. or so, as BEGEMANN et al. (1954) report all the radioactivity in now-forming cottunnite (PbCl$_2$) at fumaroles is due to ^{210}Pb. HAMAGUCHI et al. (1962) have studied ^{210}Pb and ^{212}Pb in sediments of thermal springs.

The ^{210}Pb is also useful as an atmospheric tracer (PEIRSON et al., 1966; JACOBI and ANDRÉ, 1963; MADDOCK and WILLIS, 1961; BURTON and STEWART, 1960; VILENSKII et al., 1965; and JAWOROWSKI, 1966). It is of particular interest in the dating of snow (CROZAZ and LANGWAY, 1966; CROZAZ et al., 1966; GOLDBERG, 1963) and other precipitation as again the ^{210}Pb occurs free of precursors. Some studies have used ^{210}Pb/^{206}Pb in dating of uranium minerals (BEGEMANN et al., 1952; KULP et al., 1953). PICCIOTTO (1958) and TRIPPLER (1966) have studied ^{212}Pb and ^{214}Pb in the atmosphere.

GOLDBERG (1963) has shown that the ^{210}Pb content shows a rapid depletion between the source regions of rivers (2–4 dpm/l) and downstream locations (≤ 0.1 dpm/l). The reason for the rapid depletion is not known. Major rivers in regions of major enrichments of uranium have high ^{210}Pb contents (14.7 dpm/l for Colorado River water at Grand Junction, Colorado). JURAIN (1962) has suggested that ^{210}Pb could be used in prospecting for uranium ores. GOLDBERG (1963) has shown that the turnover rate of ^{210}Pb in the oceans is very rapid, and he attributes it to biological causes. ALBERTI et al. (1959), GRANDPIERRE and ARNAUD (1965a, b) have studied ^{212}Pb and ^{214}Pb in waters.

PASTEELS (1965) has studied ^{210}Pb in zircon, and MILLARD (1963) has studied ^{212}Pb (and ^{210}Pb as measured by ^{210}Po) in zircon and other thorium-rich minerals (see U-Th-Pb Dating). VENKATASUBRAMANIAN (1963) has studied the leakage of radiolead from minerals, and NAYDENOV and CHERDYNTSEV (1958) have studied ^{210}Pb and ^{212}Pb leached from minerals.

Appendices A–D

Appendix A: *Radioactive decay schemes of uranium and thorium*
(after FRIEDLANDER and KENNEDY, 1949)

U 92	U^{238}, U_I (uranium I)		U^{234}, U_{II} (uranium II)				
Pa 91		α ・ β ・ Pa^{234}, UX_2 ・ β (99.85%) ・ I.T.(0.15%) ・ Pa^{234}, UZ ・ α						
Th 90	Th^{234}, UX_1 (uranium X_1)		Th^{230}, Io (ionium)				
Ac 89			α				
Ra 88			Ra^{226}, Ra (radium)				
Fr 87			α				
Rh 86			Rn^{222}, Rn (radon)				
At 85			α	At^{218}			
Po 84			Po^{218}, RaA (radium A)	β (0.02%) α	Po^{214}, RaC' (radium C')		Po^{210}, RaF (polonium)
Bi 83			α (99.98%)	Bi^{214}, RaC (radium C)	β (99.96%) α	Bi^{210}, RaE (radium E)	α
Pb 82			Pb^{214}, RaB (radium B)	α (0.04%)	Pb^{210}, RaD (radium D)	α $(5\times10^{-5}\%)$	Pb^{206}, RaG (stable lead isotope)
Tl 81				Tl^{210}, RaC'' (radium C'')	β	Tl^{206}, RaE'' (radium E'')	β

Appendix A: *Radioactive decay schemes of uranium and thorium (con't)*

Th 90	Th^{232}, Th (thorium)		Th^{228}, RdTh (radiothorium)		
Ac 89	α	Ac^{228}, $MsTh_2$ (mesothorium 2)	α		
		β (from Ac to Th column)			
Ra 88	Ra^{228}, $MsTh_1$ (mesothorium 1)		Ra^{224}, ThX (thorium X)		
Fr 87			α		
Rn 86			Rn^{220}, Tn (thoron)		
At 85			α	At^{216}	
Po 84			Po^{216}, ThA (thorium A)	β (0.013 %) (?) $\quad \alpha$	Po^{212}, ThC′ (thorium C′)
Bi 83			α ($\sim$100 %)	Bi^{212}, ThC (thorium C)	β (66.3 %) $\quad \alpha$
Pb 82			Pb^{212}, ThB (thorium B)	$\beta \quad \alpha$ (33.7 %)	Pb^{208}, ThD (stable lead isotope)
Tl 81				Tl^{208}, ThC″ (thorium C″)	β

Appendix A: *Radioactive decay schemes of uranium and thorium (con't)*

U 92	U^{235}, AcU (actinouranium)				
Pa 91	α	Pa^{231}, Pa (protactinium)			
Th 90	Th^{231}, UY (uranium Y)	β α	Th^{227}, RdAc (radioactinium)		
Ac 89		Ac^{227}, Ac (actinium)	β (98.8 %) α		
Ra 88		α (1.2 %)	Ra^{223}, AcX (actinium X)		
Fr 87		Fr^{223}, AcK (actinium K)	β α		
Rn 86		α (4×10^{-3} %)	Rn^{219}, An (actinon)		
At 85		At^{219}	β (3 %) α	At^{215}	
Po 84		α (97 %)	Po^{215}, AcA (actinium A)	β (5×10^{-4} %) α	Po^{211}, AcC' (actinium C')
Bi 83		Bi^{215}	β α	Bi^{211}, AcC (actinium C)	β (0.32 %) α
Pb 82			Pb^{211}, AcB (actinium B)	β α (99.68 %)	Pb^{207}, AcD (stable lead isotope)
Tl 81				Tl^{207}, AcC'' (actinium C'')	β

Appendix B: *Isotopic ratios of lead from whole-rock Precambrian and Paleozoic granites, gneisses, pegmatites, marbles, and composites of sections (see text for information on analytical techniques)*

Rock	Age	Locality	$^{206}Pb/^{204}Pb$	$^{207}Pb/^{204}Pb$	$^{208}Pb/^{204}Pb$	Reference
			Africa			
Marble	Precambrian	Bulawayan	15.01 34±1	15.19 19±1	34.37 50±1	WAMPLER and KULP (1962)
			Asia			
Aldan Shield						
Gray granite gneiss	~3000 m.y.	Baikalian block	14.10	14.87	36.78	
Gray gneiss	~3000 m.y.	Baikalian block	13.84	14.86	36.30	
Amphibolite	~3000 m.y.	Baikalian block	15.76	15.36	37.71	
Pyroxene gneiss	~3000 m.y.	Baikalian block	17.25	15.56	37.58	
Biotite gneiss	~3000 m.y.	Baikalian block	14.60	15.15	33.76	SOBOTOVICH et al. (1965)
Granite gneiss	~2700 m.y.	Baikalian block	14.65	15.09	34.61	
Amphibolite bearing migmatite	~2500 m.y.	Baikalian block	15.12	15.27	41.86	
Gray migmatite	~1700	Baikalian block	15.28	15.56	37.26	
Spotted migmatite	~1200 m.y.	Baikalian block	15.63	15.68	37.51	
Chinese-Korean Shield						
Dolomitic marble	Precambrian	Anshan, China	16.08 16.78 17.70	15.31 15.46 15.38	36.07 36.18 36.15	VINOGRADOV and TUGARINOV (1961)
Altai Mountains						
Granite		Tigerek	19.94	16.09	40.02	SOBOTOVICH and GRASHCHENKO (1965)
Granite		Tigerek	18.95	16.11	39.98	
Plagiogranite		Insk iron ore deposit	18.52	16.01	39.54	

Baltic Shield (Karelian Shield)

			Europe			
Granite gneiss	Precambrian	Pitkaranta, Karelia	26.40	17.40	46.00	VINOGRADOV and ZYKOV (1955)
Granite gneiss	Precambrian	Pitkaranta, Karelia	21.50	16.35	45.20	STARIK and SOBOTOVICH (1956)
Rapakivi Granite	Precambrian	Pitkaranta, Karelia	17.56	15.54	37.15	STARIK and SOBOTOVICH (1956)
Granite	Precambrian	Mutukhe, D. V. K.	20.50	17.10	42.25	ZHIROV and ZYKOV (1956)
Gray Granite Gneis	Precambrian	Granites of Vorona tundra, Kola Peninsula	32.98 42.32	18.41 20.08	84.96 106.01	SOBOTOVICH et al (1963c)[a]
Gray Granite Gneiss	Precambrian	Granites of Verona tundra, Kola Peninsula	15.47	15.33	36.88	
Pink Migmatite	Precambrian	Granites of Vorona tundra, Kola Peninsula	15.36	15.35	35.63	
Pink Migmatite	Precambrian	Granites of Vorona tundra, Kola Peninsula	14.92	15.16	35.36	
Quartz feldspar	Precambrian	Granites of Vorona tundra, Kola Peninsula	22.82	16.82	73.80	SOBOTOVICH et al. (1963c)[a]
Gneiss	Precambrian	Granites of Vorona tundra, Kola Peninsula	16.72	15.33	36.80	
Gneiss	Precambrian	Granites of Vorona tundra, Kola Peninsula	16.36	15.29	39.42	

[a] Ultrabasic rocks and minerals are omitted.

Appendix B

Appendix B (continued)

Europe (continued)

Rock	Age	Locality	$^{206}Pb/^{204}Pb$	$^{207}Pb/^{204}Pb$	$^{208}Pb/^{204}Pb$	Reference
Gabbro–norite	Precambrian	Monchegor pluton, Kola Peninsula	17.20	15.40	38.61	SOBOTOVICH et al. (1963c)[a]
Gabbro–norite	Precambrian	Monchegor pluton, Kola Peninsula	17.13	15.51	37.80	
Gabbro–diabase	Precambrian	Monchegor pluton, Kola Peninsula	17.20	15.40	38.80	
Gabbro–norite	Precambrian	Monchegor pluton, Kola Peninsula	17.17	15.51	37.90	
Marble	Precambrian	Kytäjä, Finland Kalkkimaa, Finland	16.26	15.39	35.84	WAMPLER and KULP (1962)
			16.00	15.53	36.04	
Gray granite	Precambrian	Quarry of Lotsman's Kamenka	16.52	15.50	36.85	
Pegmatite	Precambrian	Quarry of Lotsman's Kamenka	23.12	17.06	42.48	SOBOTOVICH et al. (1963b)
Pegmatite (Heavy fraction)	Precambrian	Quarry of Lotsman's Kamenka	29.14	17.95	50.80	
Pegmatite (Light fraction)	Precambrian	Quarry of Lotsman's Kamenka	14.87	14.98	35.20	
Porphyritic granite	Precambrian	Quarry at Korostisheva	43.75	19.06	128.50	SOBOTOVICH et al. (1963c)
Porphyritic granite	Precambrian	Quarry at Korostisheva	27.32	16.92	81.33	
Gray Granite	Precambrian	Quarry at Korostisheva	19.15	16.05	38.30	
Gray Granite	Precambrian	Quarry at Korostisheva	20.43	16.35	41.34	
Gray Granite	Precambrian	Quarry at Korostisheva	20.81	16.09	39.61	
Pink Granite	Precambrian	Adabash Kirovograd	16.12	15.62	35.85	
Gray Granite	Precambrian	Adabash Kirovograd	17.81	15.98	38.50	

Ukrainian Shield						
Gray Granite	Precambrian	Taromskoe Quarry, Ukraine (Dnieper Region)	17.20	15.50	37.70	Sobotovich et al. (1963a)
Gray Granite	Precambrian	Taromskoe Quarry, Ukraine (Dnieper Region)	16.60	15.20	36.60	
Gray Granite	Precambrian	Taromskoe Quarry, Ukraine (Dnieper Region)	15.66	15.18	36.22	
Gray Granite	Precambrian	Taromskoe Quarry, Ukraine (Dnieper Region)	15.80	15.30	37.00	
Pink Granite	Precambrian	Taromskoe Quarry, Ukraine (Dnieper Region)	14.82	15.01	34.00	
Pink Granite	Precambrian	Taromskoe Quarry, Ukraine (Dnieper Region)	17.70	15.65	38.30	Sobotovich et al. (1963a)
Gneiss	Precambrian	Taromskoe Quarry, Ukraine (Dnieper Region)	21.00	16.35	41.50	
Xenolith in gneiss	Precambrian	Taromskoe Quarry, Ukraine (Dnieper Region)	15.85	15.30	35.40	
Gray Granite	Precambrian	Iamburg Quarry	17.63	15.47	37.95	Sobotovich et al. (1963b)
Gray Granite	Precambrian	Iamburg Quarry	15.45	15.15	36.61	
Gray Granite	Precambrian	Iamburg Quarry	15.38	15.16	36.15	
Gray Granite	Precambrian	Iamburg Quarry	18.23	15.67	38.52	

Appendix B (continued)

Rock	Age	Locality	$^{206}Pb/^{204}Pb$	$^{207}Pb/^{204}Pb$	$^{208}Pb/^{204}Pb$	Reference
Europe (continued)						
Scotland: Group I (Pyroxene granulite facies)						
Leucocratic gneiss	~2600 m.y.	Scourian area	13.77	14.74	36.70	MOORBATH *et al.* (1969)
Melanocratic gneiss	~2600 m.y.	near Scoraig	15.98	15.17	36.66	
Melanocratic gneiss	~2600 m.y.	Gruinard Bay	16.15	15.23	36.50	
Leucocratic gneiss	~2600 m.y.	Torridon	13.94	14.63	34.22	
Basic lens in acid gneiss	~2600 m.y.	Torridon	15.34	15.02	36.24	
Acid gneiss	~2600 m.y.	near Lochinver	15.63	15.01	35.53	
Acid gneiss	~2600 m.y.	near Lochinver	14.69	14.97	34.92	
Basic gneiss	~2600 m.y.	near Lochinver	15.30	15.03	37.43	
Scotland: Group II (Pyroxene granulite facies retrograded to amphibolite facies)						
Basic gneiss	~2600 m.y.	near Lochinver	15.77	15.11	36.34	MOORBATH *et al.* (1969)
Leucocratic gneiss	~2600 m.y.	Gruinard Bay	13.67	14.59	45.92	
Leucocratic gneiss	~2600 m.y.	Gruinard Bay	14.32	14.70	52.11	
Leucocratic gneiss	~2600 m.y.	Gruinard Bay	13.75	14.70	39.54	
Leucocratic gneiss	~2600 m.y.	Gruinard Bay	14.27	14.79	64.87	
Leucocratic gneiss	~2600 m.y.	Gruinard Bay	14.01	14.73	34.83	
Same as above but different handspecimen	~2600 m.y.	Gruinard Bay	14.21	14.80	35.11	
Leucocratic gneiss	~2600 m.y.	Gairloch area	14.00	14.74	36.39	
Same as above but different handspecimen	~2600 m.y.	Gairloch area	14.12	14.75	37.47	
Leucocratic gneiss	~2600 m.y.	Torridon	15.88	15.06	36.50	
Leucocratic gneiss	~2600 m.y.	Torridon	14.91	14.83	37.23	

Rock type	Age	Locality				Reference
Leucocratic gneiss	~2600 m.y.	Torridon	14.10	14.69	35.36	Moorbath et al. (1969)
Mesocratic gneiss	~2600 m.y.	Torridon	15.28	14.89	47.00	
Hornblende rock from same locality as above	~2600 m.y.	Torridon	16.10	15.05	39.12	
Leucocratic gneiss	~2600 m.y.	Isle of Rona	13.77	14.60	46.17	Moorbath et al. (1969)
Grey Gneiss	~2600 m.y.	Benbecula	14.05	14.68	34.85	
Hybrid gneiss	~2600 m.y.	North Uist	14.01	14.58	33.75	

Scotland: Group III (Transition zone between pyroxene granulite facies and amphibolite facies)

Rock type	Age	Locality				Reference
Leucocratic gneiss	mixed	Claisfearn	13.50	14.74	33.92	Moorbath et al. (1969)
Mesocratic gneiss	mixed	Laxford Bridge	13.86	14.73	33.99	
Mesocratic gneiss	mixed	Badnabay	16.12	15.14	36.76	

Scotland: Group IV (Amphibolite facies superimposed on granulite facies)

Rock type	Age	Locality				Reference
Epidotised gneiss	mixed	Gleneig	13.96	14.74	35.29	Moorbath et al. (1969)
Foliated gneiss	mixed	Gleneig	13.69	14.67	35.34	
Foliated gneiss	mixed	Gleneig-Loch Hourn	16.24	15.15	41.69	
Foliated gneiss	mixed	Gleneig-Loch Hourn	14.84	14.92	40.52	

Scotland: Group V (Amphibolite facies)

Rock type	Age	Locality				Reference
Leucocratic gneiss	1600 to 1800 m.y.	Rhiconich Bridge	15.19	14.92	40.59	Moorbath et al. (1969)
Leucocratic gneiss	1600 to 1800 m.y.	near Rhiconich Bridge	15.11	14.90	41.95	
Leucocratic gneiss	1600 to 1800 m.y.	near Rhiconich Bridge	14.31	14.77	37.18	
Same as above but separate handspecimen	1600 to 1800 m.y.	near Rhiconich Bridge	14.21	14.71	37.22	
Leucocratic gneiss	1600 to 1800 m.y.	near Rhiconich Bridge	14.55	14.91	38.06	

Appendix B (continued)

Rock	Age	Locality	^{206}Pb/^{204}Pb	^{207}Pb/^{204}Pb	^{208}Pb/^{204}Pb	Reference
		North America				
Canada						
Granite	1100 m.y.	Essonville	20.25	15.65	48.73	PATTERSON (1953)
Biotite granite	1850 m.y.	S. W. Saskatchewan	25.97	16.32	45.95	ROSHOLT et al. (1970)
North America						
Granite	1450 m.y.	Uncompahgre, Colorado	20.04	15.55	36.36	PATTERSON (1953)
Granodiorite	1450 m.y.	Silver Plume, Colorado	18.08	15.67	47.33	PETERMAN et al. (1967)
Composite, Idaho Springs Formation[b]	1850 m.y.(?)	Idaho Springs, Colorado	19.48	15.77	38.74	This paper
Granite	Paleozoic	Westerly, Rhode Island	18.42	15.63	38.98	DOE et al. (1967)
Granite	1100 m.y.	Llano, Texas	18.55	15.49	38.39	
Quartz Diorite	1100 m.y.	Llano, Texas	17.93	15.51	37.76	
Gneiss	1350 m.y.(?)	Llano, Texas	23.27	15.89	41.86	ZARTMAN (1965)
Gneiss	1350 m.y.(?)	Llano, Texas	25.44	16.23	45.29	
Gneiss	1350 m.y.(?)	Llano, Texas	19.78	15.73	42.12	
Gneiss	1350 m.y.(?)	Llano, Texas	19.75	15.60	39.21	
Composite	Paleozoic	Western Montana	19.92	16.00	39.84	MURTHY and PATTERSON (1961)
Composite	Precambrian	Western Montana	18.67	15.81	39.06	
Marble	about 1100 m.y.	Canadian Shield	17.98	15.46	36.32	DOE (1962a)
			19.22	15.60	36.64	DOE (1962a)
			19.6	15.8	37.6	WAMPLER and KULP (1962)

[b] Composite was prepared by CARL E. HEDGE, U.S. Geological Survey and is composed of $^1/_3$ sillimanitic mica-plagioclase-quartz schist (metamorphosed shale), $^1/_3$ biotite-quartz-plagioclase gneiss (metagreywacke), $^1/_6$ amphibolite (metabasalt), and $^1/_6$ microcline gneiss (metadacite). The metamorphic grade of the rocks is the upper amphibolite facies.

Appendix C: *Isotope composition of lead in Phanerozoic sediments*

A: Isotopic composition of acid-leached lead in manganese nodules and pelagic sediments of the oceans

Rock	No. of samples	Locality	$^{206}Pb/^{204}Pb$	$^{207}Pb/^{204}Pb$	$^{208}Pb/^{204}Pb$	Reference
		Arctic Ocean				
Sediment	1	Arctic Ocean	19.05	15.76	39.24	CHOW and PATTERSON (1962a)
Jurassic Mn Nodule	1	Timor Island	18.78	15.84	39.33	CHOW and PATTERSON (1962a)
		Atlantic				
Sediments	7	S.W. Atlantic (E)[a]	18.82	15.82	39.22	CHOW and PATTERSON (1962a)
Mn Nodules	3	N.E. Atlantic (F)[a]	19.06	15.82	39.58	CHOW and PATTERSON (1959)
Sediments	5	N.E. Atlantic (F)[a]	19.06	15.85	39.61	CHOW and PATTERSON (1962a)
Mn Nodule	7	N.W. Atlantic (G)[a]	19.13	15.82	39.66	CHOW and PATTERSON (1959)
Sediment	17	N.W. Atlantic (G)[a]	19.11	15.82	39.65	CHOW and PATTERSON (1962a)
Sediment	1	S.E. Atlantic	19.07	15.80	39.62	CHOW and PATTERSON (1962a)
		Indian Ocean				
Sediment	1	Indian Ocean	19.21	15.86	39.90	CHOW and PATTERSON (1962a)
		Mediterranean				
Sediment	1	East central	18.42	15.69	38.25	CHOW (1968b)
		Pacific				
Red Clay	1	undesignated	18.95	15.76	38.92	PATTERSON (1953)
Sulfide bearing sediment	1	San Diego Trench	19.08	15.73	39.17	WAMPLER and KULP (1964)
Mn Nodules	2	near Japan (A)[a]	18.55	15.67	39.00	CHOW and PATTERSON (1959)
Mn Nodules	4	near Japan (A)[a]	18.51	15.62	38.80	CHOW and TATSUMOTO (1964)
Sediments	3	near Japan (A)[a]	18.36	15.61	38.51	CHOW and TATSUMOTO (1964)

[a] Letters in parentheses refer to regional designations of CHOW and PATTERSON (1962a).

Appendix C (continued)

Rock	No. of samples	Locality	206Pb/204Pb	207Pb/204Pb	208Pb/204Pb	Reference
Mn Nodules	8	North Pacific (B)[a]	18.81	15.79	39.28	CHOW and PATTERSON (1959)
Sediments	18	North Pacific (B)[a]	18.84	15.81	39.32	CHOW and PATTERSON (1962a)
Mn Nodules	10	East Pacific (C)[a]	18.85	15.78	39.29	CHOW and PATTERSON (1959)
Sediments	28	East Pacific (C)[a]	18.78	15.80	39.12	CHOW and PATTERSON (1962a)
Sediments, leaches	3	Near Guadalupe (C)[a]	18.92	15.70	38.95	CHOW, TATSUMOTO and PATTERSON (1962)
Sediments, residues	3	Near Guadalupe (C)[a]	18.97	15.65	38.93	CHOW, TATSUMOTO and PATTERSON (1962)
Mn Nodules	4	S.E. Pacific (D)[a]	18.99	15.88	39.47	CHOW and PATTERSON (1959)
Sediments	8	S.E. Pacific (D)[a]	18.94	15.82	39.38	CHOW and PATTERSON (1962a)
		Red Sea Region (from south to north)				
Sediment	2	Gulf of Aden	18.99	15.72	39.18	CHOW (1968b)
Sediment	1	southern Red Sea	18.90	15.75	38.91	CHOW (1968b)
Sediment	4	southern Red Sea	19.08	15.58	38.68	CHOW (1968b)
Sediment	2	northern Red Sea	19.02	15.62	38.64	CHOW (1968b)

B: Isotopic composition of lead in Phanerozoic whole sediments from the continents

Rock	Age	Locality	206Pb/204Pb	207Pb/204Pb	208Pb/204Pb	Type of analysis	Reference
		Africa					
Sediment	Quaternary	Gulf of Aqaba	20.25	16.05	40.31	Acid leach	CHOW (1968b)

Material	Age	Location				Type	Reference
Europe							
Shale	Paleozoic	Sweden	90.9	19.6	39.3	Whole rock	COBB and KULP (1961)
Shale	Paleozoic	Sweden	87.7	19.4	39.1	Whole rock	COBB and KULP (1961)
Shale	Paleozoic	Sweden	38.8	16.1	37.0	Whole rock	COBB and KULP (1961)
Lake sediment	Quaternary	Baltic Sea	20.57	16.00	39.70	Acid leach	CHOW (1965)
			20.72	15.98	40.40	Acid leach	CHOW (1965)
North America							
Limestone	Paleozoic	Antelope Range, Nevada	24.53	16.16	39.51	Whole rock	DOE (1970)
Limestone	Paleozoic	Antelope Range. Nevada	25.60	16.21	40.17	Whole rock	DOE (1970)
Limestone	Paleozoic	Leadville, Colorado	21.33	15.83	39.33	Whole rock	ENGEL and PATTERSON (1957)
Lomita Marl	Pleistocene	Palos Verdes Hills, Colorado	19.27	15.61	39.40	Acid soluble	PATTERSON(1953)
Mn nodule	Quaternary	Gulf of California	19.20	15.73	38.98	Acid soluble	CHOW and PATTERSON (1962a, b)
Lake sediment	Quaternary	Hudson Bay, Canada	25.00	16.41	46.83	Acid soluble	CHOW (1965)
Lake sediment	Quaternary	Hudson Bay, Canada	24.39	16.34	46.54	Acid soluble	CHOW (1965)
Lake sediment	Quaternary	Hudson Bay, Canada	24.15	16.28	45.90	Acid soluble	CHOW (1965)
Lake sediment	Quaternary	Hudson Bay, Canada	23.42	16.23	45.04	Acid soluble	CHOW (1965)
Lake sediment	Quaternary	Hudson Bay, Canada	23.49	16.35	45.01	Acid soluble	CHOW (1965)
Lake sediment	Quaternary	Hudson Bay, Canada	22.36	16.17	44.66	Acid soluble	CHOW (1965)
Lake sediment	Quaternary	Hudson Bay, Canada	22.07	16.12	42.71	Acid soluble	CHOW (1965)
Lake sediment	Quaternary	Hudson Bay, Canada	21.63	16.00	43.37	Acid soluble	CHOW (1965)
Lake sediment	Quaternary	Great Slave Lake, Canada	23.41	16.38	49.33	Acid soluble	CHOW (1965)
			20.09	15.97	40.69	Acid soluble	CHOW (1965)
Lake sediment	Quaternary	Great Bear Lake, Canada	24.15	16.59	44.10	Acid soluble	CHOW (1965)
			22.08	16.18	42.13	Acid soluble	CHOW (1965)
			21.56	16.13	41.77	Acid soluble	CHOW (1965)
			21.46	16.15	41.37	Acid soluble	CHOW (1965)

Appendix C (continued)

Rock	Age	Locality	$^{206}Pb/^{204}Pb$	$^{207}Pb/^{204}Pb$	$^{208}Pb/^{204}Pb$	Type of analysis	Reference
Lake sediment	Quaternary	Lake Superior, N. America	20.53	15.87	40.34	Whole rock	HART and TILTON (1966)
			22.84	16.34	42.86	Acid leach	HART and TILTON (1966)
			22.48	16.20	42.34	Water leach	HART and TILTON (1966)
Sand	Recent	Salton Sea, California	19.40	15.75	39.30	Acid soluble	DOE, HEDGE, and WHITE (1966)
			18.55	15.64	38.51	Residue	DOE, HEDGE, and WHITE (1966)
Claystone	Recent	Salton Sea, California	19.40	15.75	39.33	Acid soluble	DOE, HEDGE, and WHITE (1966)
			19.28	15.69	39.14	Residue	DOE, HEDGE, and WHITE (1966)
Claystone	Pliocene	Salton Sea, California	19.35	15.75	39.32	Acid soluble	DOE, HEDGE, and WHITE (1966)
			19.41	15.75	39.41	Residue	DOE, HEDGE, and WHITE (1966)
Clastone	Pliocene	Slaton Sea, California	19.18	-15.67	39.06	Acid soluble	DOE, HEDGE, and WHITE (1966)
			19.29	15.68	39.16	Residue	DOE, HEDGE, and WHITE (1966)
Silty claystone	Pliocene or Miocene	Salton Sea, California	19.13	15.67	39.01	Acid soluble	DOE, HEDGE, and WHITE (1966)
			18.99	15.68	39.23	Residue	DOE, HEDGE, and WHITE (1966)
Sandstone	Pliocene	Salton Sea, California	17.90	15.55	37.67	Residue	MUFFLER and DOE (1968)
Sandstone	Pliocene or Miocene	Salton Sea, California	17.95	15.57	37.72	Residue	MUFFLER and DOE (1968)
Siltstone	Cenozoic	Salton Sea, California	19.22	15.67	39.13	Residue	MUFFLER and DOE (1968)
Lithic wacke (basaltic sandstone)	Eocene	Oregon	19.81	15.76	39.77	Whole rock	TATSUMOTO and SNAVELY (1969)
Arkosic or Lithic wacke	Eocene	Oregon	19.22	15.82	39.65	Whole rock	TATSUMOTO and SNAVELY (1969)
Silty sandstone	Oligocene	Oregon	19.15	15.74	39.27	Whole rock	TATSUMOTO and SNAVELY (1969)

Appendix D. Isotopic composition of lead in Cenozoic and Cretaceous igneous rocks

Description	$^{206}Pb/^{204}Pb$	$^{207}Pb/^{204}Pb$	$^{208}Pb/^{204}Pb$	References
A. Ocean Basins				
Abyssal basalt, Mid-Atlantic Ridge	*Atlantic Ocean Basin*			
High alumina tholeiite	18.47	15.54	38.01	TATSUMOTO (1966 b)
High alumina tholeiite	17.82	15.54	37.52	TATSUMOTO (1966 b)
High alumina tholeiite	18.82	15.68	38.65	TATSUMOTO (1966 b)
Island volcanics, Mid-Atlantic Ridge				
Ascension Island				
Olivine basalt	19.43	15.67	39.20	GAST *et al.* (1964) [a]
Olivine-poor basalt	19.55	15.68	39.04	GAST (1967) [a]
Syenite inclusion	19.58	15.70	39.38	GAST (1967)
Trachyte (avg. of 3 samples)	19.71	15.71	39.44	GAST (1967) [a]
Obsidian bomb	19.50	15.64	39.21	GAST (1967) [a]
St. Helena				
Picrite basalt	20.98	15.93	40.49	GAST (1969) [a]
Trachybasalt	20.91	15.87	40.44	
Phonolite (avg. of 5 samples)	20.925	15.891	40.514	
Gough Island				
Olivine-poor basalt	18.37	15.68	38.98	GAST *et al.* (1964) [a]
Porphyritic trachybasalt	18.43	15.74	39.26	
Porphyritic trachyandesite	18.39	15.69	39.15	
Trachyte (avg. of 2)	18.67	15.74	39.55	

[a] Analyses were made on Ta (lot 1) filament material which appeared to have no correction for mass spectrometer bias. Ratios should therefore be close to absolute.

Appendix D (continued)

Description	$^{206}Pb/^{204}Pb$	$^{207}Pb/^{204}Pb$	$^{208}Pb/^{204}Pb$	References
Iceland, Group I				WELKE et al. (1968)
Olivine basalt (avg. of 2)	18.41	15.63	38.55	
Dacite (avg. of 2)	18.55	15.64	38.52	
Granophyre (avg. of 4)	18.43	15.57	38.42	
Iceland, Group II				
Olivine basalt (avg. of 3)	18.95	15.63	38.85	
Andesite	18.89	15.65	38.82	
Felsite	18.82	15.61	38.70	
Obsidian (avg. of 2)	18.92	15.61	38.78	
Granophyre (avg. of 3)	19.03	15.67	38.97	
Open Ocean				
Vema Seamount				
Phonolite from Collins Peak	19.82	15.88	40.05	COOPER and RICHARDS (1966c)
Pacific Ocean Basin				
Abyssal basalts, East Pacific Rise				
High alumina tholeiites	18.19	15.54	37.93	TATSUMOTO (1966b)
High alumina tholeiites	18.24	15.53	38.03	TATSUMOTO (1966b)
High alumina tholeiites	18.50	15.58	38.34	TATSUMOTO (1966b)
Island volcanics, East Pacific Rise				
Easter Island				
Obsidian, Mount Ourito	19.31	15.67	39.17	PATTERSON and DUFFIELD (1963)
Obsidian, Mount Ourito	19.31	15.66	39.15	TATSUMOTO (1966b)
Andesine andesite	19.25	15.58	38.94	TATSUMOTO (1966b)
Alkali basalt	19.30	15.73	39.46	TATSUMOTO (1966b)
Tholeiite with alkalic affinities	19.28	15.67	39.16	TATSUMOTO (1966b)

Guadalupe Island				
Labradorite-andesine alkali basalt	20.44	15.74	40.78	TATSUMOTO (1966b)
Labradorite-andesine alkali basalt	20.28	15.73	40.55	TATSUMOTO (1966b)
Labradorite alkali basalt	20.17	15.76	40.49	TATSUMOTO (1966b)
Labradorite olivine basalt with alkalic affinities	20.18	15.67	40.31	TATSUMOTO (1966b)
Volcanic rich sediment				
Feldspar	19.33	15.66	39.13	PATTERSON and TATSUMOTO (1964)
Iwo Jima, Volcano Islands				
Trachyandesite	19.56	15.76	39.51	TATSUMOTO (1966a)
Open, ocean, Hawaiian Islands				
Oahu				
Tholeiitic basalt	18.09	15.58	38.24	TATSUMOTO (1966a)
Melilite-nepheline basalt (avg of 2)	18.20	15.69	38.25	TATSUMOTO (1966a)
Alkali olivine basalt, Upper Waianae series	18.03	15.48	37.78	COOPER and RICHARDS (1966b)
Hawaiite, Upper Waianae series	18.21	15.65	38.35	COOPER and RICHARDS (1966b)
Tholeiitic basalt, Middle Waianae series	18.14	15.55	37.98	COOPER and RICHARDS (1966b)
Olivine tholeiitic basalt, Lower Waianae series	17.98	15.57	37.95	COOPER and RICHARDS (1966b)
Rhyodacite, Mauna Kuwale volcano	17.87	15.53	37.90	PATTERSON (1964)
Hawaii				
Alkaliolivine basalt, Hualalai volcano	18.03	15.56	38.01	PATTERSON (1964)
Anorthite from feldspathic bomb	18.05	15.51	37.86	PATTERSON (1964)
Alkaliolivine basalt, Hualalai volcano	17.94	15.51	37.85	TATSUMOTO (1966a)
Hawaiite, Mauna Kea volcano	18.47	15.63	38.40	TATSUMOTO (1966a)
Ankaramite, Mauna Kea volcano	18.46	15.58	38.30	TATSUMOTO (1966a)
Trachyte, Hualalai volcano	18.08	15.49	38.22	TATSUMOTO (1966a)
Trachyte, Hualalai volcano	18.00	15.50	37.86	PATTERSON (1964)
Trachyte, Kohala Mtn.	18.50	15.68	38.38	TATSUMOTO (1966a)
Maui				
Mugearite (avg. of 2) and trachyte, Honolua series	18.60	15.69	38.65	COOPER and RICHARDS (1966b)

Appendix D (continued)

Description	$^{206}Pb/^{204}Pb$	$^{207}Pb/^{204}Pb$	$^{208}Pb/^{204}Pb$	References
Olivine basalt (avg. of 3), Wailuku series	18.44	15.62	38.30	COOPER and RICHARDS (1966 b)
Molokai				
Olivine basalt, Eeast Molokai series	18.35	15.53	38.01	COOPER and RICHARDS (1966 b)
Mugearite, East Molokai series	18.39	15.51	38.06	COOPER and RICHARDS (1966 b)
Basalt, West Molokai series	18.00	15.54	37.94	COOPER and RICHARDS (1966 b)
Kauai				
Nepheline basalt, Kolea series	18.36	15.66	38.54	COOPER and RICHARDS (1966 b)
Tholeiitic basalt, Makaweli formation (avg. of 2)	18.03	15.58	38.05	COOPER and RICHARDS (1966 b)
Hawaiite, Makaweli formatiòn	18.43	15.68	38.55	COOPER and RICHARDS (1966 b)
Tholeiitic basalt, Napali formation	18.19	15.56	38.06	COOPER and RICHARDS (1966 b)
Indian Ocean Basin				
Olivine basalt, Reunion Island (avg. of 2)	18.53	15.69	38.77	COOPER and RICHARDS (1966 b)
B. Island Arcs				
Japan				
Tholeiitic (pigeonitic or calcic), high alumina, and andesitic (hypershenic or calc-alkalic) series (from south to north)				
Oshima				
Olivine tholeiite, Okata basalt group Pliocene	18.45	15.69	38.69	TATSUMOTO and KNIGHT (1969)
"Autointrusion" in the above*	18.54	15.67	38.77	TATSUMOTO and KNIGHT (1969)
"Autointrusion" in the above*	18.41	15.62	38.49	COOPER and RICHARDS (1966 b)
Olivine tholeiite, Mihara-yama, probably older than 4000 B.C.	18.49	15.68	38.72	TATSUMOTO and KNIGHT (1969)

* The value of $^{206}Pb/^{204}Pb$ of the "autointrusion" appears to be different from that of the host basalt. As no effect is noted in $^{207}Pb/^{204}Pb$ or $^{208}Pb/^{204}Pb$, neither instrumental nor natural isotopic fractionation can be involved. Some more complex process must be involved than mere "autointrusion". Either the source magma was isotopically heterogeneous (or assimilating lead rapidly from some other source), or, as TATSUMOTO and KNIGHT suggest, volatile transfer of lead from an isotopically different source was occurring.

Olivine augite tholeiite, Mihara-yama, 4000 B.C. (?)	18.52	15.68	38.78	TATSUMOTO and KNIGHT (1969)
Olivine augite tholeiite, Mihara-yama, 4000 B.C. (?)	18.40	15.68	38.66	COOPER and RICHARDS (1966b)
Tholeiite, 1552 A.D.	18.54	15.70	38.80	TATSUMOTO and KNIGHT (1969)
Pyroxene tholeiite, 1778 A.D.	18.55	15.72	38.80	TATSUMOTO and KNIGHT (1969)
Pyroxene tholeiite, 1951 A.D.	18.53	15.68	38.77	TATSUMOTO (1966a)
Pyroxene tholeiite, Mihara-yama, 1951 A.D.	18.67	15.93	38.62	MASUDA (1964)**

Hakone

Hornblende-hypersthene dacite, upper Miocene (high aluminum series)	18.25	15.81	38.29	MASUDA (1964)**
Olivine-pyroxene tholeiite, early Pliocene	18.47	15.68	38.75	TATSUMOTO and KNIGHT (1969)
Olivine-pyroxene tholeiite, late Pliocene	18.46	15.68	38.75	TATSUMOTO and KNIGHT (1969)
Same locality as above	18.45	15.64	38.60	TATSUMOTO (1966a)
Tholeiite, old somma lava, older Pleistocene	18.33	15.63	38.54	TATSUMOTO and KNIGHT (1969)
Pyroxene andesite, old somma lava, older Pleistocene (avg. of 3), tholeiite series	18.40	15.67	38.68	TATSUMOTO and KNIGHT (1969)
Pyroxene dacite, old somma lava	18.42	15.95	38.42	MASUDA (1964)**
Pyroxene dacte, Pleistocene	18.47	15.81	38.47	MASUDA (1964)**
Two pyroxene andesite and pyroxene dacite, young somma lava (avg. of 2), tholeiite series	18.39	15.65	38.63	TATSUMOTO and KNIGHT (1969)
Two pyroxene andesite, young somma lava (avg. of 2), tholeiite series	18.49	15.86	38.41	MASUDA (1964)**
Two pyroxene andesite, central cone (avg. of 2), high alumina series	18.39	15.67	38.68	TATSUMOTO and KNIGHT (1969)
Two pyroxene andesite and hypersthene dacite, age uncertain, high aluminum series	18.47	15.74	38.30	MASUDA (1964)**

** While the pioneering work of MASUDA (1964) is valuable, the data should be used with care along with other data. Apparently he had a small contamination peak under ^{207}Pb in his analyses as his values of ^{207}Pb/^{204}Pb are clearly greater than those of others.

Appendix D (continued)

Description	$^{206}Pb/^{204}Pb$	$^{207}Pb/^{204}Pb$	$^{208}Pb/^{204}Pb$	References
Amagi-Omuroyama				
Two pyroxine andesite, late Pleistocene	18.38	15.70	38.75	Tatsumoto and Knight (1969)
Primitive high alumina basalt, Recent	18.35	15.64	38.57	Tatsumoto and Knight (1969)
Primitive high alumina basalt, Recent	18.16	15.61	38.30	Cooper and Richards (1966b)
Olivine two-pyroxene dacite, Recent (avg. of 2 samples)	18.34	15.67	38.65	Tatsumoto and Knight (1969)
Hornblende-pyroxene dacite, Pleistocene (avg. of 2 samples)	18.32	15.76	38.19	Masuda (1964)**
Two pyroxene andesite, Recent	18.35	15.65	38.56	Tatsumoto and Knight (1969)
High alumina basalt, Recent	18.37	15.67	38.66	Tatsumoto and Knight (1969)
Olivine andesite, Recent	18.32	15.67	38.61	Tatsumoto and Knight (1969)
Xenocryst-rich part of above	18.34	15.66	38.62	Tatsumoto and Knight (1969)***
Olivine andesite, older than 2400 yr.	18.32	15.65	38.58	Tatsumoto and Knight (1969)
Fuji-san or Fuji-yama				
(This is the classic volcano with the catenary sides. Tholeiitic volcanes are more flat, alkalic volcanoes are more rough.)				
High alumina basalt, old Fujii volcano	18.42	15.69	38.75	Tatsumoto and Knight (1969)
High alumina basalt, Mishima flow, prehistoric, Shimotokari quarry	18.39	15.66	38.66	Tatsumoto and Knight (1969)
Aphyric andesite segregation vein from above Shimotokari quarry	18.45	15.73	38.86	Tatsumoto and Knight (1969)****
High alumina basalt, Mishima flow, prehistoric, Shimotokari quarry	18.49	15.74	38.96	Cooper and Richards (1966b)

*** Note that this obviously highly contaminated andesite shows no isotopic departures from its apparently little contaminated comparion.

**** The lead isotopic composition of this segregation vein departs from its parent in almost a perfect reverse fractionation pattern of 0.16 percent enrichment per mass unit of the heavy isotope. The authors were aware of instrumental fractionation and made an attempt to circumvent it. However, the effect appears to be too large for natural processes

	18.xx	15.xx	38.xx	
High alumina basalt, probably erupted within the last 2000 or 3000 years	18.40	15.66	38.66	Tatsumoto and Knight (1969)
Two pyroxene basalt, 864 A.D.	18.42	15.70	38.79	Tatsumoto and Knight (1969)
Iwate, northern Honshu				
Tholeiitic basalt	18.63	15.74	39.03	Hedge and Knight (1969)
Andesite (tholeiitic series)	18.65	15.72	39.03	Hedge and Knight (1969)
Moriyosi-yama, northern Honshu				
Andesite, high alumina series	18.67	15.74	39.05	Hedge and Knight (1969)
Dacite, high alumina series	18.63	15.69	38.93	Hedge and Knight (1969)
Showa-shinzan, Hokkaido				
Dacite, 1949 eruption	18.71	15.76	39.16	Somayajulu et al. (1966)
Alkalic series				
Karatsu-Takashima, Kyushu				
Olivine trachybasalt (avg. of 2)	18.06	15.72	38.08	Masuda (1964)**
Titanaugite-olivine trachybasalt	17.80	15.46	38.09	Kurasawa (1968)
Ontake, central Honshu				
Pyroxene andesite	18.48	15.75	39.23	Tatsumoto (1969)
Yatsuga-take, central Honshu				
Pyroxene andesite	18.49	15.72	38.97	Tatsumoto (1969)
Hamada, southern Honshu				
Nepheline basalt	18.31	15.76	38.71	Masuda (1964)**
Oki-Dogo, west of Honshu				
Olivine basalt, Recent, Saigo Bay	17.98	15.74	37.89	Masuda (1964)**
Olivine basalt, Recent, Saigo Bay	18.13	15.69	38.90	Cooper and Richards (1966b)
Olivine basalt, Recent, Saigo Bay (avg. of 4 samples)	18.09	15.56	38.60	Kurasawa (1968)
Trachyte, Pliocene dike, Saigo Bay	18.21	15.79	38.74	Masuda (1964)**
Titanaugite olivine basalt	17.95	15.58	38.66	Kurasawa (1968)
Titanaugite-olivine trachy-basalt	17.86	15.53	38.39	Kurasawa (1968)

Appendix D (continued)

Description	$^{206}Pb/^{204}Pb$	$^{207}Pb/^{204}Pb$	$^{208}Pb/^{204}Pb$	References
Titanaugite-plagioclase mugearite	17.81	15.55	38.58	KURASAWA (1968)
Mugearite	17.71	15.58	38.58	KURASAWA (1968)
Hypersthene-augite plagioclase augite	18.28	15.53	38.51	KURASAWA (1968)
Olivine hedenbergite trachyte	17.80	15.60	38.72	KURASAWA (1968)
Hypersthene-titanaugite trachyte	17.93	15.61	38.83	KURASAWA (1968)
Rhyolite obsidian	17.86	15.56	38.64	KURASAWA (1968)
Rhyolite	17.97	15.57	38.84	KURASAWA (1968)
Sanidine rhyolite	17.92	15.56	38.71	KURASAWA (1968)
Atsumi				
Olivine dolerite	18.05	15.79	38.13	MASUDA (1964)**
Kampu-zan, Ichinome Gata				
Andesite (2samples)	18.65	15.68	38.85	HEDGE and KNIGHT (1969)
Rhyodacite	18.50	15.69	38.89	HEDGE and KNIGHT (1969)
Gabbro inclusion	18.39	15.66	38.72	HEDGE and KNIGHT (1969)
Granite inclusion	18.51	15.70	38.86	HEDGE and KNIGHT (1969)
Oshima-Oshima				
Basalt	18.32	15.68	38.60	HEDGE and KNIGHT (1969)
Andesite	18.45	15.64	38.57	HEDGE and KNIGHT (1969)
New Zealand				
Coromandel				
First period andesites, propylitized (avg. of 3)	18.78	15.63	38.67	COOPER and RICHARDS (1969) b [b]
Second period andesites, unaltered (avg. of 2)	18.82	15.61	38.64	
Te Aroha				
First period andesites, propylitized	18.79	15.63	38.69	
First period andesites, propylitized	18.51	15.61	38.42	
Second period andesites, unaltered (avg. of 2)	18.69	15.57	38.47	

Third period rhyolites:				
Older series	18.83	15.62	38.70	Oversby and Gast (1968)
Younger series (avg. of 2)	18.78	15.65	38.68	
C. Mediterranean Volcanics				
Italy				
Vesuvio				
Nepheline basalt	19.14	15.78	39.48	Houtermans et al. (1964)
Cotunnite, avg. of 1872, 1907, and 8 August 1952	19.08	15.74	39.31	Houtermans et al. (1964)
Cotunnite, avg. of 24 August 1952, 16 March 1954 and 16 April 1954	19.17	15.78	39.48	Houtermans et al. (1964)
Mt. Volcano				
Cannizarite (avg. of 2)	19.48	15.82	39.73	Houtermans et al. (1964)
Mt. Somma				
Galena	19.10	15.73	39.20	Eberhardt et al. (1962)
Isola d'Ischia Forio	18.99	15.78	39.43	Eberhardt et al. (1962)
D. Continental Igneous Rocks				
Africa				
Nigeria				
Panian basalt, Recent (?)	20.22	16.19	39.91	Tugarinov et al. (1968)
South Africa				
Kimberlite, Kimberley pipe (calculated initial ratios), age is 174×10^6 yrs.	19.51	15.82	39.32	Lovering and Tatsumoto (1968)
Australia				
New South Wales				
Basaltic nephelinite, Delegate pipe (calculated initial ratios) age is 68×10^6 yrs.	18.30	15.69	38.31	Lovering and Tatsumoto (1968)

Appendix D (continued)

Description	$^{206}Pb/^{204}Pb$	$^{207}Pb/^{204}Pb$	$^{208}Pb/^{204}Pb$	References
Western Victoria				
Basanite of Newe Volcanics, Mt. Leura	18.54	15.61	38.68	COOPER and GREEN (1969)[b]
Basanite of Newer Volcanics, Mt. Shadwell	18.43	15.54	38.46	COOPER and GREEN (1969)[b]
Basanite of Newer Volcanics, Mt. Noorat	18.52	15.56	38.51	COOPER and GREEN (1969)[b]
Basanite of Newer Volcanics, Mt. Gambier	18.35	15.51	38.44	COOPER and GREEN (1969)[b]
Basanite of Newer Volcanics, Mt. Schank	18.38	15.61	38.61	COOPER and GREEN (1969)[b]
Europe				
Rockall Bank, granite (avg. of 2)	17.11	15.34	37.16	MOORBATH and WELKE (1968a)
Skye, Scotland				
Lock Ainort epigranite	15.78	15.34	36.78	HAMILTON (1966)
leucocratic separate	15.60	15.30	36.82	
leucocratic separate	15.41	15.08	36.71	
S. Porphyritic epigranite	15.50	15.11	36.38	
Glas Beinn Mohr epigranite	16.60	15.40	37.48	
Pitchstone	16.61	15.61	38.12	
Pitchstone, Glas Beinn Mohr	16.68	15.88	39.48	
Lundy granite	18.87	15.47	38.21	
Ailsa Craig granite	17.14	15.47	38.31	
Skye, Scotland				
Olivine basalt of Beinn Edra Group, Trotternish	16.99	15.44	37.10	MOORBATH and WELKE (1969)
Olivine basalt of Beinn Edra Group, Portree, resting on Jurassic sediments	16.02	15.12	36.63	
Olivine basalt of Beinn Edra Group, Trotternish	16.29	15.25	37.32	
Olivine-poor, feldspar-pyric basalt of Osdale Group, Bracadale	17.73	15.56	37.80	

[b] Mass spectrometer bias removed through use of double spiking technique calibrated by gravimetric standards.

Mugearite of Beinn Totaig Group, near Portree	16.16	15.20	36.55
Olivine dolerite of the Lealt sill, Trotternish	16.82	15.38	37.16
Dolerite of the Cone-sheet, Cuillins Centre	17.90	15.56	37.96
Peridotite from dike intruding basalts, near Sligachan	17.86	15.58	38.02
Main peridotite of the Cuillins layered intrusion, Loch Coruisk	18.13	15.66	38.44
Gabbro series of zone IV of Cuillins layed intrusion, Druim an Eidhne	17.60	15.55	38.00
Leucocratic band from banded rocks from zone I of the Eucrite Series of the Cuillins layered intrusion	17.84	15.69	37.97
Melanocratic band from above specimen	17.77	15.67	37.84
Porphyritic felsite, W. Redhills Centre	15.12	14.95	36.97
Another specimen of above	15.12	14.90	36.49
Porphyritic felsite, W. Redhills Centre	15.27	15.01	36.33
Porphyritic felsite, W. Redhills Centre	15.28	14.97	36.13
Granophyre of Loch Ainort Granite, W. Redhills Centre	15.37	15.04	36.46
Granophyre from near above sample	15.15	14.99	36.23
Glamaig Granite, W. Redhills Centre	15.68	15.09	36.51
Beinn an Dubhaich Granite, E. Redhills Centre	15.95	15.15	37.69
Granite from near above sample	16.00	15.22	37.81
Granite from near above two samples	16.02	15.15	37.67
Granophyre from Glas Bheinn Mohr-Dunan Granite, E. Redhills Centre	16.50	15.37	37.84
Dolerite from Cnoc Carnach sill, near Broadford, slightly acidified, olivine-free, basic bottom member	16.03	15.15	37.42
Felsite from Cnoc Carnach sill	15.91	15.16	37.52
Dolerite from Rudh'an Eireannaich sill, Broadford, slightly acidified, olivine-free basic top member	16.41	15.28	37.48

Appendix D (continued)

Description	$^{206}Pb/^{204}Pb$	$^{207}Pb/^{204}Pb$	$^{208}Pb/^{204}Pb$	References
Dolerite from bottom member of above	16.39	15.28	37.59	
Felsite, slightly basified central acid member of Rudh'an Eireannaich sill	16.40	15.28	37.57	
Ferrodiorite from Harker's Gully, W. Redhills Centre	16.28	15.20	37.48	
Marscoite from Harker's Gully, W. Redhills Centre	15.32	15.05	37.04	
Lewisian basement rock xenolith in ferrodiorite, Harker's Gully, W. Redhills Centre, an oligoclase-quartz-biotite gneiss, recrystallized at grain boundaries	15.73	15.14	36.97	
East Greenland				
Skaergaard intrusion				HAMILTON (1966)
Acid granophyre with pyrrhotite	16.86	15.48	38.29	
Ferrogabbro with sulfides	17.48	15.18	36.39	
North America				
West Coast of North America (on or west of the Quartz Diorite Line of Moore)				
Alaska				
K-feldspar, granite of Hume Creek, Seward Peninsula	18.98	15.72	38.93	This paper (Table 19)[c]
K-feldspar, granite of Brooks Mtns., Seward Peninsula	19.09	15.72	39.13	This paper (Table 19)[c]
California				
Obsidian, Salton Sea (avg. of 3)	18.90	15.62	38.60	DOE et al. (1966)[a]
K-feldspar, Leucogranite, southern Calif. batholith	18.95	15.62	38.52	BANKS and SILVER (1964)

[c] Data are normalized to agree with those from Ta (lot 1) filament material and therefore should be close to absolute ratios.

	206Pb/204Pb	207Pb/204Pb	208Pb/204Pb	
Vasquez formation, rhyodacrite, Los Angeles	19.16	15.70	39.24	Doe (1968)[a]
Obsidian, Clear Lake	19.16	15.64	38.87	Doe (1967)[a]
Obsidian, Little Glass Mtn.	18.98	15.69	38.93	Doe (1967)[a]
Oregon				
Obsidian, John Day area	19.22	15.71	39.42	Doe (1967)[a]
Tholeiitic basalt, Coffin Butte, early to middle Eocene	18.98	15.61	38.74	Tatsumoto and Snavely (1969)[d]
Alkalic basalt, Widow Creek, early to middle Eocene (avg. of 2)	19.34	15.70	39.32	Tatsumoto and Snavely (1969)[d]
Augite basalt, Siletz River, early to middle Eocene	19.41	15.78	39.68	Tatsumoto and Snavely (1969)[d]
Gabbro, Silitz River, early to middle Eocene	19.26	15.67	39.16	Tatsumoto and Snavely (1969)[d]
Picrite basalt, early to middle Eocene	19.28	15.58	39.02	Tatsumoto and Snavely (1969)[d]
Feldspar-phyric basalt, Ball Mtn., early to middle Eocene	19.56	15.70	39.53	Tatsumoto and Snavely (1969)[d]
Porphyritic basalt, Cape Mtn., upper Eocene	19.23	15.65	39.00	Tatsumoto and Snavely (1969)[d]
Dacite, Devil's Churn, upper Eocene	19.34	15.74	39.44	Tatsumoto and Snavely (1969)[d]
Nepheline syenite, Table Mtn., upper Eocene	19.59	15.78	39.57	Tatsumoto and Snavely (1969)[d]
Biotite camptonite, Siletz River, upper Eocene	20.05	15.81	40.07	Tatsumoto and Snavely (1969)[d]
Granophyric gabbro, Lambert Point, upper Eocene	19.20	15.75	39.21	Tatsumoto and Snavely (1969)[d]
Basalt, chilled border from Mary's Peak sill, upper Eocene*	19.31	15.75	39.34	Tatsumoto and Snavely (1969)[d]
Granophyric diorite, from Mary's Peak sill*	19.22	15.74	39.30	Tatsumoto and Snavely (1969)[d]
Pegmatite, from Mary's Peak sill*	19.25	15.75	39.26	Tatsumoto and Snavely (1969)[d]
Aplite dike, from Mary's Peak sill*	19.11	15.73	39.30	Tatsumoto and Snavely (1969)[d]

[d] Calculated initial ratios.

* The present day values of $^{206}Pb/^{204}Pb$ of the aplite dike is 19.51 and 19.25 for the granophyric diorite; however, the aplite dike lead is less radiogenic than that of the granophyric diorite after correction for decay of contained uranium. If the sill is younger than supposed or if the U/Pb has been altered recently, the apparent variation of 1% in the ratio could be easily accounted for. As the authors state, all ratios are within the overall analytical uncertainties.

Appendix D (continued)

Description	$^{206}Pb/^{204}Pb$	$^{207}Pb/^{204}Pb$	$^{208}Pb/^{204}Pb$	References
Diorite, Nehkahnie Mtn., Oregon, middle Miocene	18.83	15.80	39.52	TATSUMOTO and SNAVELY (1969)[d]
Tholeiitic basalt, Depoe Bay, Oregon, middle Miocene	18.80	15.79	39.36	TATSUMOTO and SNAVELY (1969)[d]
Tholeiitic basalt, Cape Foulweather, Oregon middle Miocene	18.97	15.73	39.13	TATSUMOTO and SNAVELY (1969)[d]
Washington				
Tholeiitic basalt, Alder Bluff, Washington middle Miocene	18.48	15.78	40.14	TATSUMOTO and SNAVELY (1969)[d]
Tholeiitic basalt, Pack Sack Look-out, Washington, middle Miocene	18.35	15.72	39.92	TATSUMOTO and SNAVELY (1969)[d]
Obsidian, Mount Rainier, Washington	19.10	15.68	39.03	DOE (1967)[c]
Continental Interior of North America (east of Quartz Diorite line of Moore)				
California				
Obsidian, Mono Craters	19.12	15.66	38.92	DOE (1968)[c]
Colorado				
K-feldspar, Eldora stock Front Range (avg. of 2)	17.86	15.49	38.53	DOE and HART (1963)[a]
Basalt, Dotsero crater, contaminated	17.93	15.63	38.63	DOE *et al.* (1969 b)[c]
K-feldspar, Albion stock, Front Range	18.09	15.53	39.26	DOE and HART (1963)[a]
Vitrophyre, Willow Creek, San Juan Mountains	18.74	15.63	38.09	DOE (1967)[c]
Vitrophyre, Fisher Quartz Latite	18.36	15.58	37.68	DOE (1967)[c]
Basalt, Hinsdale basalt, Jarosa Mesa, slightly contaminated	18.38	15.62	38.26	DOE (1967)[c]
Basalt, Hinsdale basalt, Race Creek, contaminated	18.84	15.56	37.98	DOE *et al.* (1969 b)[c]

Description				Reference
Basalt, Hinsdale basalt, La Jara Reservoir region, very contaminated	17.89	15.52	37.29	Doe et al. (1969 b)[c]
Plagioclase from Hinsdale basalt sample above, light density fraction	17.98	15.60	37.63	Doe et al. (1969 b)[c]
Plagioclase from Hinsdale basalt sample above, heavy density fraction	18.27	15.58	37.91	Doe et al. (1969 b)[c]
Basalt, Hinsdale basalt, La Jara Reservoir region, slightly contaminated	17.96	15.56	37.64	Doe et al. (1969 b)[c]
Basalt. Hinsdale basalt. La Jara Reservoir region. uncontaminated	18.33	15.55	37.67	Doe et al. (1969 b)[c]
Alkalic andesite, Conejos Formation, Conejos Peak area	18.59	15.66	38.04	Doe et al. (1969a)[c]
Alkalic andesite, second sample from above locality	18.18	15.63	37.87	Doe et al. (1969a)[c]
Rhyolite dike, Conejos Formation, Summer Coon volcanic center	17.53	15.52	37.14	Doe et al. (1969a)[c]
Rhyodacite core, Conejos Formation, Summer Coon volcanic center	17.38	15.54	37.15	Doe et al. (1969a)[c]
Basalt dike, Conejos Formation, Summer Coon volcanic center	17.47	15.59	37.41	Doe et al. (1969a)[c]
Idaho				
Vitrophyre, Walcott Tuff, American Falls	18.36	15.76	38.80	Doe (1967)[c]
Basalt, primitive, Shoshone Falls	18.12	15.45	38.08	Patterson et al. (1955)
Basalt, primitive, Shoshone Falls	18.35	15.62	38.88	Tilton in Doe (1968)
Basalt, contaminated, Indian Tunnel flow, Craters of the Moon	17.97	15.69	38.84	Doe (1968)[c]
Highway andesite, contaminated, Craters of the Moon	17.81	15.60	38.66	Doe (1968)[c]
Montana				
Pre-batholic volcanics				
Biotitic vitrophyre, Elkhorn Mountains Volcanics (calculated initial ratios)	18.00	15.61	38.22	Doe et al. (1968)[c]

Appendix D (continued)

Description	$^{206}Pb/^{204}Pb$	$^{207}Pb/^{204}Pb$	$^{208}Pb/^{204}Pb$	References
Boulder batholith				
K-feldspar, Butte Quartz Monzonite and related rocks (avg. of 3)	17.95	15.58	38.19	MURTHY and PATTERSON (1961)
K-feldspar, Butte Quartz Monzonite and related rocks	18.24	15.75	38.86	MURTHY and PATTERSON (1961)
K-feldspar, Butte Quartz Monzonite and related rocks, Kain Quarry	17.99	15.60	38.23	DOE et al. (1968)[c]
Whole-rock, Butte Quartz Monzonite (calculated initial values)	18.01	15.62	38.22	DOE et al. (1968)[c]
Range (avg. of 12 samples) K-feldspars from Butte Quartz Monzonite and related rocks	17.87–18.17 (17.98)	15.54–15.65 (15.60)	38.11–38.46 (38.23)	DOE et al. (1968)[c]
K-feldspars, Ringing Rocks pluton, Boulder batholith				
Quartz monzonite core	17.96	15.62	38.38	DOE et al. (1968)[c]
Mafic monzonite rim	17.80	15.60	38.38	DOE et al. (1968)[c]
K-feldspar, syenogabbro of Kokoruda Ranch Complex	18.35	15.67	38.22	DOE et al. (1968)[c]
K-feldspars, Unionville Granodiorite and Burton Park pluton (avg. of 5 samples)	18.02	15.61	38.28	DOE et al. (1968)[c]
K-feldspar, aplite dike in Unionville Granodiorite	17.89	15.64	38.40	DOE et al. (1968)[c]
K-feldspars, Rader Creek granodiorite pluton (avg. of 3 samples)	16.94	15.44	37.74	DOE et al. (1968)[c]
K-feldspar, pegmatite mass in Rader Creek pluton	16.85	15.46	37.76	DOE et al. (1968)[c]
K-feldspar, Hell Canyon pluton (avg. of 2 samples)	17.72	15.56	38.54	DOE et al. (1968)[c]

K-feldspar, aplite sheet in Hell Canyon pluton (avg. of 2 analyses)	17.54	15.57	38.61	Doe et al. (1968)[c]
K-feldspar, equigranular facies of Donald pluton and pegmatite (avg. of 2 samples)	17.39	15.62	38.47	Doe et al. (1968)[c]
K-feldspar, mainly groundmass of strongly porphyritic facies of Donald pluton (avg. of two analyses)	17.00	15.45	37.95	Doe et al. (1968)[c]
K-feldspar (groundmass and megacryst)	17.01	15.49	38.12	Doe et al. (1968)[a]
Plagioclase (groundmass)	17.12	15.60	38.49	Doe et al. (1968)[a]
K-feldspar (megacryst)	17.19	15.51	38.13	Doe et al. (1968)[a]
Post-batholith volcanics				
K-feldspar. post-Lowland Creek volcanics. post-batholith	18.57	15.68	38.54	Doe et al. (1968)[a]
Plagioclase, post-batholith Lowland Creek Volcanics	18.15	15.64	38.64	Doe et al. (1968)[a]
Garnet Range *				This paper
Contaminated basalt	17.27	15.41	37.07	
Slightly contaminated basalt	17.86	15.47	38.02	

* This pair of basalts are thought to be possibly genetically related highly contaminated (64 P-9) (46°38′ N, 112°36′ W) and slightly contaminated (64 P-94) (46°51′ N, 113°05′ W) basalts from the Garnet Range, Montana. The samples were donated by H. J. Prostka and described by P. W. Lipman, both of the U.S. Geological Survey, Denver, Colorado. The slightly contaminated basalt contains about 15% coarse grained phases. Of these, olivine comprises about 30%, clinopyroxene about 30% in the form of crystal aggregates, 20% opaque oxides and 20% plagioclase with many grains sieved with inclusions or have irregularly bounded cores of more sodic plagioclase rimmed by more calcic plagioclase which are in turn zoned outward to more sodic material. The zoned plagioclases are probably indicative of partially melted xenocrysts. The groundmass is comprised of plagioclase (50%), clinopyroxene (25%), olivine (10%), opaque oxides (5%), apatite (1%) and interstitial material (9%). The slightly contaminated basalt is a fairly typical basalt, except that the biotite in the interstitial material indicates unusual residual concentration of potassium (see chemical analysis). The highly contaminated basalt contains about 30% coarse grained phases. Of these, biotite rimmed with probably cognate opaques comprises 45%, clinopyroxene about 30% (several large grains, intergrown in a clot of large apatites, are probably xenocrystic), plagioclase about 3% and probably xenocrystic), plagioclase about 3% and probably xenocrysts, olivine and altered olivine about 15%, quartz xenocrysts about 6% and apatite xenocrysts about 1%. The groundmass is comprised of plagioclase (50%), sodic plagioclase (?) (15%), clinopyroxene (15%), opaque oxides (10%), and interstitial material (10%). This basalt is not only clearly contaminated but also a peculiar rock in general which before contamination must have been a rather mafic Fe- and K-rich rock as indicated by abundant biotite, the groundmass opaques and some K-feldspar. This rock is called a lamphrophyre by H. J. Prostka.

Rapid Rock Analysis (by Leonard Shapiro, *project leader)*

	Slightly contaminated basalt	Highly contaminated basalt
SiO_2	50.0	52.9
Al_2O_3	17.4	14.6
Fe_2O_3	2.4	3.9
FeO	5.8	2.8
MgO	5.8	5.4
CaO	8.6	8.0
Na_2O	2.7	2.6
K_2O	1.8	3.9
H_2O^-	1.1	1.1
H_2O^+	1.7	1.3
TiO_2	1.8	1.4
P_2O_5	0.44	1.2
MnO	0.14	0.13
CO_2	0.05	0.21
Totals	100	99

Appendix D (continued)

Description	$^{206}Pb/^{204}Pb$	$^{207}Pb/^{204}Pb$	$^{208}Pb/^{204}Pb$	References
Nevada				
K-feldspar, rhyolite vitrophyre Busted Butte section	18.18	15.69	39.42	This paper (Table 19)[c]
K-feldspar, quartz latite pumice	18.23	15.66	39.29	This paper (Table 19)[c]
K-feldspar, rhyolite vitrophyre, Ammonia tanks, Pahute Mesa	18.11	15.68	39.24	This paper (Table 19)[c]
K-feldspar, quartz latite dev., Ammonia tanks, Pahute Mesa	18.41	15.72	39.30	This paper (Table 19)[c]
K-feldspar, quartz porphyry at Rock House	19.38	15.75	39.28	Doe (1968)[c]
K-feldspar, quartz porphyry at Cortez gold locality	19.39	15.75	39.28	Doe (1968)[c]
K-feldspar, Caetano tuff	19.31	15.72	39.11	Doe (1968)[c]

New Mexico				
Olivine basalt, primitive, Jemez Mtns. (avg. of 2 analyses)	18.34	15.46	37.86	Doe (1967)[c]
Olivine basalt, contaminated, Jemez Mtns. (avg. of 2 analyses)	17.78	15.50	37.60	Doe (1967)[c]
Obsidian, Arroyo Hondo, Jemez Mtns.	18.20	15.53	38.00	Doe (1967)[c]
Obsidian, Los Posos, Jemez Mtns.	18.01	15.57	37.99	Doe (1967)[c]
Obsidian from Castle Knob near Silver City	18.25	15.56	38.14	Doe (1967)[c]
Basalt of the Servilleta formation of Montgomery (1953), from Taos area (avg. of 2), contaminated	17.38	15.50	37.21	Doe et al. (1969b)[c]
Uncontaminated basalt from above locality	18.09	15.51	37.55	Doe et al. (1969b)[c]
South Dakota				
Pitchstone, Black Hills	17.93	15.66	38.54	Doe et al. (1967b)[c]
Texas				
Rhyolite vitrophyre, Tascotal Mesa	17.80	15.65	37.94	Doe (1967)[c]
Soda rhyolite, Tascotal Mesa	17.81	15.66	38.10	Doe (1967)[c]
Utah				
K-feldspar, granite of North Star Range	18.41	15.65	38.69	Stacey et al. (1968)
K-feldspar, Blue Star pluton, Mineral Range	18.76	15.70	38.28	Stacey et al. (1968)
K-feldspar, granite of Desert Mountain	18.86	15.71	39.18	Stacey et al. (1968)
K-feldspar, Silver City pluton, Tintic area	18.68	15.71	38.96	Stacey et al. (1968)
K-feldspar, monzonite of Bingham Canyon (avg. of 2)	17.83	15.70	38.72	Stacey et al. (1968)
K-feldspar, soda syenite porphyry, La Sal Mtns.	19.27	15.65	38.60	Stern et al. (1965)
K-feldspar, monzonite porphyry, La Sal Mtns.	19.37	15.72	38.69	Stern et al. (1965)
K-feldspar, diorite porphyry, La Sal Mtns.	18.27	15.55	37.62	Stern et al. (1965)
Washington				
Obsidian, Bodie Mtn.	19.00	15.67	38.84	Doe (1967)[c]

Appendix D (continued)

Description	$^{206}Pb/^{204}Pb$	$^{207}Pb/^{204}Pb$	$^{208}Pb/^{204}Pb$	References
Wyoming				
Obsidian, Cougar Creek, Yellowstone National Park	16.58	15.44	38.58	DOE (1967)
Obsidian, Gibbon River, Yellowstone National Park	17.18	15.59	38.40	DOE (1967)
Shoshonite from Sunlight crater, corrected to initial values	16.86	15.44	37.32	PETERMAN *et al.* (1970)
Rhyodacite welded tuff from near Sunlight volcanic center, feldspar	16.88	15.50	37.64	PETERMAN *et al.* (1970)
Shoshonite of Elk Creek basalt, corrected to initial values	17.06	15.56	37.94	PETERMAN *et al.* (1970)
Rhyodycite welded tuff from near the above locality, feldspar	17.30	15.58	37.64	PETERMAN *et al.* (1970)
Biotite-hornblende andesite from the Electric Peak volcanic center, corrected to initial values	16.62	15.45	37.33	PETERMAN *et al.* (1970)
Pyroxene-hornblende andesite from the Washburn volcanic center, corrected to initial values	16.31	15.45	36.94	PETERMAN *et al.* (1970)
Pyroxene-hornblende andesite from the Sunlight volcanic center, corrected to initial values	16.47	15.38	36.82	PETERMAN *et al.* (1970)

References

AKISHIN, P. A., NIKITIN, O. T., PANCHENKOV, G. M.: A new effective ion emitter for the isotopic lead analysis. Geokhimiya, no. 5, 429—434 (1967).

ALBERTI, G., BETTINALI, C., SALVETTI, F.: Behavior and determination of RaB, RaD, and ThB in natural waters. Ann. Chim. (Rome) **49**, 193—198 (1959).

ALDRICH, L. T., TILTON, G. R., DAVIS, G. L., NICOLAYSEN, L. O., PATTERSON, C. C.: Comparison of U-Pb, Pb-Pb, and Rb-Sr ages of Precambrian minerals. In: DERRY, D. R.: Symposium on Precambrian Correlation and Dating. Can. Geol. Assoc. Proc. **7**, pt. 2, 7—13 (1955).

— DAVIS, G. L., TILTON, G. R., WETHERILL, G. W.: Radioactive ages of minerals from the Brown Derby mine and the Quartz Creek granite near Gunnison, Colorado. J. Geophys. Res. **61**, no. 2, pt. 1, 215—232 (1956).

— WETHERILL, G. W.: Geochronology by radioactive decay. In: SEGRÈ, E. G. (ed.), Annual review of nuclear science. Natl. Res. Council, Natl. Acad. Sci., Stanford, Calif., Ann. Rev. **8**, 257—298 (1958).

— — TILTON, G. R., DAVIS, G. L.: Half-life of Rb[87]. Phys. Rev. **103**, 1045—1047 (1956a).

ALLEGRE, CLAUDE, J., LANCELOT JOEL: A new method for the mass spectrometric isotope analysis of lead contained in rocks and minerals (in French). Acad. Sci. Paris Compt. Rend. Ser. D **266**, 1646—1648 (1968).

ANDREYEV, P. F., BUGROV, N. M., GLEBOVSKAYA, V. S., DANILOV, L. T., IL'INSKAYA, Y. A., KASHTAN, M. S., KESHISHYAN, G. O.: Isotopic composition of lead in natural waters. Geokhimiya, no. 6, 761—676 (1967)

ARMSTRONG, R. L.: A model for the evolution of strontium and lead isotopes in a dynamic earth. Rev. Geophys. **6**, 175—199 (1968).

ASTON, F. W.: The constitution of ordinary lead. Nature **120**, 224 (1927).

— Mass spectrum of uranium lead and the atomic weight of protactinium. Nature **123**, 313 (1929).

BANKS, P. O., SILVER, L. T.: Material balance in the whole rock U-Pb system of a young granite. Am. Geophys. Union Trans. **45**, 108 (1964).

— — Evaluation of the decay constant of uranium 235 from lead isotopic ratios. J. Geophys. Res. **71**, 4037—4046 (1966).

BEGEMANN, F., VON BUTTLER, H., HOUTERMANS, F. G., ISAAC, N., PICCIOTTO, E. E.: Preliminary results of age measurements of Shinkolobwe pitchblende by the RaD method. Soc. Belge Géol. Bull. **61**, pt. 2, 223—226 (1952).

— GEISS, J., HOUTERMANS, F. G., BUSER, W.: Isotopic composition and radioactivity of recent Vesuvius lead. Nuovo Cimento **11**, 663—673 (1954).

BERMAN, R. M.: The role of lead and excess oxygen in uraninite. Am. Mineral. **42**, 705—731 (1957).

BHANOT, V. B., JOHNSON, W. H., JR., NIER, A. O.: Atomic masses in the heavy mass region. Phys. Rev. **120**, 235—251 (1960).

BROWN, J. S.: Oceanic lead isotopes and ore genesis. Econ. Geol. **60**, pp. 47—68 (1965).

— Isotopic zoning of lead and sulfur in southeast Missouri. In: Genesis of stratiform lead-zinc-fluorite deposits (Mississippi Valley type deposits). A symposium, New York, 1966. Econ. Geol., Mon. 3, 410—425 (1967).

BURGER, A. J., NICOLAYSEN, L. O., DE VILLIERS, J. W. L.: Lead isotopic compositions of galenas from the Witwatersrand and Orange Free State, and their relation to the Witwatersrand and Dominion Reef uraninites. Geochim. Cosmochim. Acta **26**, 25—59 (1962).
— VON KNORRING, O., CLIFFORD, T. N.: Mineralogical and radiometric studies of monazite and sphene occurrences in the Namib Desert, South-West Africa. Mineral. Mag. **35**, 519—528 (1965).
— NICOLAYSEN, L. O., AHRENS, L. H.: Controlled leaching of monazites. J. Geophys. Res. **75**, 3585—3594 (1967).
BURKSER, E. S., YELISEYEVA, G. D., LECHEKHLEV, V. R., SHCHERBAK, N. P.: Migration of lead in monazite and pitchblende. Akad. Nauk SSSR Kom. Opredeleniyu Absolyut. Vozrasta Geol. Formatsii Byull., no. 5, 48—52 (1962).
BURTON, W. M., STEWART, N. G.: Use of long-lived natural radioactivity as an atmospheric tracer. Nature **186**, 584—589 (1960).
CAHEN, L., EBERHARDT, P., GEISS, J., HOUTERMANS, F. G., JEDWAB, J., SIGNER, P.: On a correlation between the common lead model age and the trace-element content of galenas. Geochim. Cosmochim. Acta **14**, 134—149 (1958).
CAMERON, A. E., SMITH, D. H., WALKER, R. L.: Mass spectrometry of nanogram-size samples of lead. Anal. Chem. **41**, 525—526 (1969).
CANNON, R. S., JR., PIERCE, A. P., ANTWEILER, J. C.: The data of lead isotope geology related to problems of ore genesis. Econ. Geol. **56**, 1—38 (1961).
— STEIFF, L. R., STERN, T. W.: Radiogenic lead in nonradiogenic minerals: A clue in the search for uranium and thorium. Proc. Intern. Conf. Peaceful Uses At. Energy (Geneva) **2**, 215—220 (1958).
CATANZARO, E. J.: Zircon ages in southwestern Minnesota. J. Geophys. Res. **68**, 2045—2048 (1963).
— Triple-filament method for solid-sample lead isotope analysis. J. Geophys. Res. **72**, 1325—1327 (1967).
— The interpretation of zircon ages, pp. 225—258. In: HAMILTON, E. I., FARQUHAR, R. M. (Eds.): Radiometric dating for geologists, p. 506. New York: Interscience Publ. 1968.
— GAST, P. W.: Isotopic composition of lead in pegmatitic feldspars. Geochim. Cosmochim. Acta **19**, 113—126 (1960).
— KULP, J. L.: Discordant zircons from the Little Belt (Montana), Beartooth (Montana) and Santa Catalina (Arizona) Mountains. Geochim. Cosmochim. Acta **28**, 87—124 (1964).
— MURPHY, T. J., SHIELDS, W. R., GARNER, E. L.: Absolute isotopic abundance ratios of common, equal-atom and radiogenic lead isotope standards. J. Res. Natl. Bur. Std. A. Phys. Chem. **72**A, 261—267 (1968).
CHOUBERT, B.: Absolute age of Precambrian rocks of Guiana (in French). Acad. Sci. Paris. Compt. Rend. **258**, 631—634 (1964).
CHOW, T, J.: Radiogenic leads of the Canadian and Baltic Shield regions. In: Symposium on Marine Geochemistry, 1964. Rhode Island Univ. Narragansett Marine Lab. Occasional Pub. **3**, 169—184 (1965).
— Isotope analysis of seawater by mass spectrometry. J. Water Pollution Control Federation **40**, pt. 1, 399—411 (1968a).
— Lead isotopes of the Red Sea region. Earth Planet. Sci. Letters **5**, 143—147 (1968b).
— JOHNSTONE, M. S.: Lead isotopes in gasoline and aerosols of Los Angeles basin, California. Science **147**, 502—503 (1965).
— PATTERSON, C. C.: Lead isotopes in manganese nodules. Geochim. Cosmochim. Acta **17**, 21—31 (1959).

— — On the primordial lead of the Canyon Diablo meteorite. Geokhimiya, no. 12, 1124—1125 (1961).

— — The occurrence and significance of lead isotopes in pelagic sediments. Geochim. Cosmochim. Acta **26**, 263—308 (1962a).

— — Correction [to "Lead isotopes in manganese nodules"]. Geochim. Cosmochim. Acta **26**, 973 (1962b).

— TATSUMOTO, M.: Isotopic composition of lead in the sediments near Japan Trench, Chap. 10 (179—184). In: Recent Researches in the Fields of Hydrosphere, Atmosphere, and Nuclear Geochemistry, 404 p. Tokyo: Maruzen Co. Ltd. 1964.

CLIFFORD, T. N., NICOLAYSEN, L. O., BURGER, A. J.: Petrology and age of the pre-Otavi basement granite at Franzfontein, northern South-West Africa. J. Petrol. **3**, 244—278 (1962).

COBB, J. C.: Dating of black shales, pp. 311—316. In: Geochronology of rock systems. N. Y. Acad. Sci. Ann. **91**, 594 p. (1961).

— Determination of lead in meteorites by alpha activation analysis. J. Geophys. Res. **69**, 1895—1901 (1964).

COLLINS, C. B., FREEMAN, J. R., WILSON, J. T.: A modification of the isotopic lead method for determination of geological ages. Phys. Rev. **82**, 966—967 (1951).

COMPSTON, W., OVERSBY, V. M.: Lead isotopic analysis using a double spike. J. Geophys. Res. **74**, 4338—4348 (1969).

COOPER, J. A., GREEN, D. H.: Lead isotope measurements on lherzolite inclusions and host basanites from western Victoria, Australia. Earth Planet. Sci. Letters **6**, 69—76 (1969).

— RICHARDS, J. R.: Solid-source lead isotope measurements and isotopic fractionation: Earth Planet. Sci. Letters **1**, 58—64 (1966a).

— — Lead isotopes and volcanic magmas. Earth Planet. Sci. Letters **1**, 259—269 (1966b).

— — Isotopic and alkali measurements from the Vema Seamount of the South Atlantic Ocean. Nature **210**, 1245—1246 (1966c).

— — Lead isotope measurements on sediments from Atlantis II and Discovery deep areas, pp. 499—511. In: Hot brines and recent heavy metal deposits in the Red Sea, 600 p. Berlin-Göttingen-Heidelberg: Springer (1969a).

— — Lead isotope measurements on volcanics and associated galenas from Coromandel-Te Aroha region, New Zealand. Geochem. J. (Japan) **3**, 1—14 (1969b).

CROZAZ, G., LANGWAY, C. C., JR.: Dating Greenland firn-ice cores with Pb-210. Earth Planet. Sci. Letters **1**, 194—196 (1966).

— — PICCIOTTO, E. E.: Artificial radioactivity reference horizons in Greenland Firn. Earth Planet. Sci. Letters **1**, 42—48 (1966).

DARNLEY, A. G.: Uranium-thorium-lead age determinations with respect to the Phanerozoic time-scale, pp. 73—86. In: The Phanerozoic time-scale. 548 p. Geol. Soc. London Quart. J. **120**, Suppl., (1964).

DAVIS, G. L., HART, S. R., TILTON, G. R.: Some effects of contact metamorphism on zircon ages. Earth Planet. Sci. Letters **5**, 27—34 (1968).

DELEVAUX, M. H.: Lead reference sample for isotopic abundance ratios. In: Short papers in geology and hydrology. U.S. Geol. Survey Prof. Paper 475-B, B160-B161 (1963).

— DOE, B. R., BROWN, G. F.: Preliminary lead isotope investigations of brine from the Red Sea, galena from the Kingdom of Saudi Arabia, and galena from the United Arab Republic (Egypt). Earth Planet. Sci. Letters **3**, 139—144 (1967).

DoE, B. R.: Distribution and composition of sulfide minerals at Balmat. New York, Geol. Soc. Am. Bull. **73**, 833—854 (1962a).
— Relationships of lead isotopes among granites, pegmatites, and sulfide ores near Balmat. New York. J. Geophys. Res. **67**, 2895—2906 (1962b).
— The bearing of lead isotopes on the source of granitic magma. J. Petrol. **8**, 51—83 (1967).
— Lead and strontium isotope studies of Cenozoic volcanic rocks in the Rocky Mountain region — a summary. Colorado School Mines Quart. **63**, 149—174 (1968a).
— A list of references on lead isotope geochemistry through 1966. U.S. Geol. Survey open-file report, 97 p. (1968b).
— Evaluation of U-Th-Pb whole-rock dating on Phanerozoic sedimentary rocks. Eclogae Geological Helvetica **63**, in press (1970).
— HEDGE, C. E., WHITE, D. E.: Preliminary investigation of the source of lead and strontium in deep geothermal brines underlying the Salton Sea geothermal area. Econ. Geol. **61**, 462—483 (1966).
— LIPMAN, P. W., HEDGE, C. E.: Radiogenic tracers and the source of continental andesites; a beginning at the San Juan volcanic field, Colorado. Proc. Andesites Conf., Internat. Upper Mantle Sci. Rept. **16**, State Oregon Dept. Geol. Mineral. Ind. Bull. **65**, 143—149 (1969a).
— — — KURASAWA, H.: Primitive and contaminated basalts from the Southern Rocky Mountains, U.S.A. Contrib. Mineral. Petrol. **21**, 142—156 (1969b).
— NEWELL, M. F.: Isotopic composition of uranium in zircon. Am. Mineral. **50**, 613—618 (1965).
— TATSUMOTO, M., DELEVAUX, M. H., PETERMAN, Z. E.: Isotope-dilution determination of five elements in G-2 (granite), with a discussion of the analysis of lead. In: Geological Survey research 1967. U.S. Geol. Survey Prof. Paper 575-B, B170-B177 (1967).
— TILLING, R. I., HEDGE, C. E., KLEPPER, M. R.: Lead and strontium isotope studies of the Boulder batholith, southwestern Montana. Econ. Geol. **63**, 884—906 (1968).
— TILTON, G. R., HOPSON, C. A.: Lead isotopes in feldspars from selected granitic rocks associated with regional metamorphism. J. Geophys. Res. **70**, 1947—1968 (1965).
EBERHARDT, P., GEISS, J., HOUTERMANS, F. G.: Lead and sulfur isotope ratios in galenas. Helv. Phys. Acta **28**, 339—341 (1955).
— — — SIGNER, P.: Age determinations on lead ores. Geol. Rundschau **52**, 836—852 (1962).
ENGEL, A. E. J., PATTERSON, C. C.: Isotopic composition of lead in Leadville limestone, hydrothermal dolomite, and associated ore. Geol. Soc. Amer. Bull. **68**, 1723 (1957).
FARQUHAR, R. M., RUSSELL, R. D.: Anomalous leads from the upper Great Lakes region of Ontario. Am. Geophys. Union Trans. **38**, 552—556 (1957).
FLEMING, E. H., JR., GHIORSO, A., CUNNINGHAM, B. B.: The specific alpha-activities and half-lives of U^{234}, U^{235}, and U^{236}. Phys. Rev. **88**, 642—652 (1952).
FRIEDLANDER, G., KENNEDY, J. W.: Nuclear and radiochemistry. New York: John Wiley and Sons, Inc., 468 p. (1955).
GAST, P. W.: Isotope geochemistry of volcanic rocks, pp. 325—358. In: Basalts — The Poldervaart treatise on rocks of basaltic composition, Vol. **1**, 482 p. New York: Interscience Publ. 1967.
— The isotopic composition of lead from St. Helena and Ascension Islands. Earth Planet. Sci. Letters **5**, 353—359 (1969).

— TILTON, G. R., HEDGE, C. E.: Isotopic composition of lead and strontium from Ascension and Gough Islands. Science **145**, 1181—1185 (1964).

GEISS, J.: Isotope analyses of "common lead". Z. Naturforsch. **9a**, 218—227 (1954).

GERLING, E. K., ISKANDEROVA, A. D.: Isotopic composition of lead from carbonate rocks of different age. Akad. Nauk SSSR Doklady **170**, 905—907 (1966).

GOLDBERG, E. D.: Geochronology with lead-210. In: Symposium on Radioactive Dating, Athens, 1962, Proc. Vienna, Internat. Atomic Energy Agency, 121—131 (1963).

GOLUBCHINA, M. N.: Determination of the absolute age of phosphorite by the lead-isotopic method. Inform. Sb., Vses. Nauchno-Issled. Geol. Inst., no. 54, 27—29 (1962).

GRAESER, VON, S., HUNZIKER, J. C.: Rb-Sr and Pb isotope compositions in rocks and minerals from the Ivrea zone. Schweiz. Mineral. Petrog. Mitt. **48**, 189—204 (1968).

GRANDPIERRE, R., ARNAUD, M.: Spectrometric investigations of the absorption of natural radioactive elements taken at the Luchon thermal resort. Ann. Inst. Hydrol. Climatol. **33**, 40—51 (1965a).

— ARNAUD, Y.: The conditions of the therapeutic employment of radioactive mineral water. Presse Thermale Climat. **102**, 167—171 (1965b).

GRÜNENFELDER, M. H.: Heterogeneity of accessory zircons and the petrographic significance of their uranium decay age; Pt. 1, Zircons of granodiorite gneisses of Acquacalda (Lukmanier Pass). Schweiz. Mineral. Petrog. Mitt. **43**, 235—257 (1963).

— HANSON, G. N., BRUNNER, G. O., EBERHARD, E.: U-Pb discordance and phase unmixing in zircons. In: Abstracts for 1966. Geol. Soc. America Spec. Paper 101, p. 80 (1968).

— HOFMANNER, F., GROGLER, N.: Heterogeneity of accessory zircons, and the petrographic significance of their uranium-lead decay age; Pt. 2, Precambrian zircon formation in the Gotthard massif. Schweiz. Mineral. Petrog. Mitt. **44**, 543—558 (1964).

GUNTEN, H. R. VON, WYTTENBACH, A., DULAKAS, H.: Half-life of ^{210}Pb(RaD). J. Inorg. Nucl. Chem. **29**, 2826—2829 (1967).

HAMAGUCHI, H., LI, Y. T., CHENG, H. S.: The radioactivity of hokutolite. J. Chinese Chem. Soc. **9**, 1 —13 (1962).

HAMILTON, E. I.: Applied geochronology, with a chapter on comparative geochemistry. London-New York: Academic Press, 267 p., 1965.

— The isotopic composition of lead in igneous rocks — Pt. 1, The origin of some Tertiary granites. Earth Planet. Sci. Letters **1**, 30—37 (1966).

HART, S. R., DAVIS, G. L., STEIGER, R. H., TILTON, G. R.: A comparison of the isotopic mineral age variations and petrologic changes induced by contact metamorphism, pp. 73—110. In: HAMILTON, E. I., FARQUHAR, R. M. (Eds.). Radiometric dating for geologists, 506 p. New York: Interscience Publ. 1968.

— KROGH, T. E., DAVIS, G. L., ALDRICH, L. T., MUNIZAGA, F.: Rb/Sr geochronology of granitic rocks southeast of Sudbury, Ontario. Carnegie Inst. Wash. Year Book **65**, 1965—66, 59—63 (1967).

— TILTON, G. R.: The isotope geochemistry of strontium and lead in Lake Superior sediments and water, pp. 127—137. In: The earth beneath the continents, 663 p. Am. Geophys. Union Geophys. Mon. Ser., no. **10** (1966).

HEDGE, C. E., KNIGHT, R. J.: A study of lead and strontium isotopes in volcanic rocks from northern Honshu, Japan. Geochem. J. **3**, 15—24 (1969).

HEIMLICH, R. A., BANKS, P. O.: Radiometric age determinations, Bighorn Mountains, Wyoming. Am. J. Sci. **266**, 180—192 (1968).

HEYL, A. V., DELEVAUX, M. H., ZARTMAN, R. E., BROCK, M. R.: Isotopic study of galenas from the upper Mississippi Valley, the Illinois-Kentucky, and some Appalachian Valley mineral districts. Econ. Geol. **61**, 933—961 (1966).

HOLMES, A., CAHEN, L.: African geochronology 1956; results to July 1, 1956. Acad. Roy. Sci. Col., Cl. Sci. Nat. Mém. in −8°, n.s., **5**, 169 p. (1957).

HORNE, J. E. T., DAVIDSON, C. F.: The age of the mineralization of the Witwatersrand. G. Brit. Geol. Survey Bull. 10, 58—73 (1955).

HOUTERMANS, F. G., EBERHARDT, A., FERRARA, G.: Lead of volcanic origin, Chap. 18, pp. 233—243. In: Isotopic and cosmic chemistry, 553 p. Amsterdam: North-Holland Publ. Co. 1964.

INGHRAM, M. G.: Manhattan Project, Tech. Ser., National Nuclear Energy Ser., Div. II, Vol. **14**, Chap. 5, p. 35. New York: McGraw-Hill Book Co. 1946.

IORDANOV, N.: Allanite study and determination of absolute geological age of Plana pluton. Akad. Nauk SSSR Kom. Opredeleniyu Absolyut. Vozrasta Geol. Formatsii Trudy, sess. **7**, 274—282 (1960).

ISKANDEROVA, A. D.: Preliminary data on determination of the absolute age of carbonate sediments by the method of ordinary lead. Akad. Nauk SSSR Kom. Opredeleniyu Absolyut. Vozrasta Geol. Formatsii Trudy, sess. **13**, 449—455 (1966).

— LEGIERSKIY, Y.: Use of apatite for determination of the absolute age of geological formations by the lead-isotope method. Akad. Nauk SSSR Kom. Opredeleniyu Absolyut. Vozrasta Geol. Formatsii Trudy, sess. **13**, 444—448 (1966). (Also see Geophys. Abstracts, October 1967.)

IVANTISHIN, M. M., ALEKSEEVA, K. N., DEMIDENKO, S. G., ELISEEVA, G. D., KOTLOVSKAYA, F. I.: New data on the age of the Korosten pluton determined by lead and rubidium methods. Akad. Nauk SSSR Kom. Opredeleniyu Absolyut. Vozrasta Geol. Formatsii Trudy, sess. **10**, 105—111 (1961).

JACOBI, W., ANDRÉ, K.: The vertical distribution of radon 222, radon 220, and their decay products in the atmosphere. J. Geophys. Res. **68**, 3799—3814 (1963).

JAWOROWSKI, Z.: Temporal and geographical distribution of radium D (lead-210). Nature **212**, 886—889 (1966).

JURAIN, G.: Contribution to the geochemical knowledge of the families of uranium-radium and of thorium in the southern Vosges — Application of some results to prospecting of deposits of uranium. Sci de la Terre Mem., no. 1, 349 p. (1962).

KANASEWICH, E. R.: Approximate age of tectonic activity using anomalous lead isotopes. Roy. Astron. Soc. Geophys. J. **7**, 158—168 (1962).

— The interpretation of lead isotopes and their geological significance, pp. 147—223. In: HAMILTON, E. I., FARQUHAR, R. M. (Eds.). Radiometric dating for geologists, 506 p. New York: Interscience Publ. 1968a.

— Precambrian rift; genesis of strata-bound ore deposits. Science **161**, 1002—1005 (1968b).

— FARQUHAR, R. M.: Lead isotope ratios from the Cobalt-Noranda area, Canada. Can. J. Earth Sci. **2**, 361—384 (1965).

— SLAWSON, W. F.: Precision intercomparisons of lead isotope ratios, Ivigtut, Greenland. Geochim. Cosmochim. Acta **28**, 541—549 (1964).

KAUTZSCH, E., BIRKENFELD, H., ZAHN, H., CHANG, I., KAEMMEL, T., KRUHME, H.: Lead isotopic abundance of the lead ores of East Germany. Deut. Akad. Wiss. Berlin. Kl. Chemie, Geologie, Biologie, no. 7, 865—876 (1964).

KOEPPEL, V.: Age and history of the uranium mineralization of the Beaverlodge area, Saskatchewan. Can. Geol. Survey Paper **67**—**31**, 111 p. (1968).

KOLLAR, F., RUSSELL, R. D., ULRYCH, T. J.: Precision intercomparisons of lead isotope ratios, Broken Hill and Mount Isa. Nature **187**, 754—756 (1960).

KOMLEV, L. V., IVANOVA, K. S., SAVONENKOV, V. G.: On the differential mobility of lead isotopes and character of the admixed lead in monazites (in Russian, with English summ.). Geokhimiya, no. 12, 1228—1239 (1964).

— MIKHALEVSKAYA, A. D., DANILEVICH, S. I.: The age of the alkaline intrusions of the Chibina and Lovozero tundras (Kola Peninsula). Akad. Nauk SSSR Doklady **136**, 172—174 (1961).

KOSZTOLANYI, C.: New method of isotope analysis of zircons in natural state (in French). Centre Rech. Radiogeol., Nancy, France, Compt. Rend. **260**, no. 22, Groupe 9, 5849—5851 (1965).

KOUVO, O.: Radioactive age of some Finnish pre-Cambrian minerals. Finlande Comm. Geol. Bull. **182**, 70 p. (1958).

KOVARIK, A. F., ADAMS, N. I., JR.: The disintegration constant of thorium and the branching ratio of thorium C. Phys. Rev. **54**, 413—421 (1938).

— — Redetermination of the disintegration constant of U^{238}. Phys. Rev. **98**, no. 1, p. 46 (1955).

KULP, J. L., BROECKER, W. S., ECKELMANN, W. R.: Age determination of uranium minerals by the Pb^{210} method. Nucleonics **11**, 19—21 (1953).

KURASAWA, H.: Isotopic composition of lead and concentrations of uranium, thorium, and lead in volcanic rocks from Dōgo of the Oki Islands, Japan. Geochem. J. **2**, 11—28 (1968).

LAMBERT, I. B., HEIER, K. S.: Geochemical investigations of deep-seated rocks in the Australian shield. Lithos **1**, 30—53 (1968).

LARSEN, E. S., JR., KEEVIL, N. B., HARRISON, H. C.: Method for determining the age of igneous rocks using the accessory minerals. Geol. Soc. Am. Bull. **63**, 1045—1052 (1952).

LIEBENBERG, W. R.: The occurrence and origin of gold and radioactive minerals in the Witwatersrand system, the Dominion reef, the Ventersdorp contact reef and the Black reef (with discussion). Geol. Soc. South Africa Trans. **58**, 105—254 (1955).

LOUW, J. D.: Geological age determinations on Witwatersrand uraninites using the lead isotope method (with an appendix on Chemical analyses, by STRELOW, F. W. E.). Geol. Soc. South Africa Trans. **57**, 209—230 (1954).

LOVERING, J. F., TATSUMOTO, M.: Lead isotopes and the origin of granulite and eclogite inclusions in deep-seated pipes. Earth Planet. Sci. Letters **4**, 350—356 (1968).

LYAKHOVICH, V. V.: Accessory minerals and the absolute age of igneous rocks. Trudy, Inst. Mineralogii, Geokhimii i Kristallokhim. Redkikh Elementov, no. 7, 212—225 (1961).

MADDOCK, A. G., WILLIS, E. H.: Atmospheric activities and dating procedures. In: Advances in Inorg. Chem. and Radiochem., Vol. 3, pp. 287—335. New York: Academic Press 1961.

MANTON, W. I., TATSUMOTO, M.: Isotopic composition of lead and strontium in eclogites and related rocks. Annual Report Geosci. Div., 1967—1968, Southwest Center Advanc. Stud. 19—20, 22 (1969).

MARSHALL, R. R.: Lead-lead age of the Bondoc meteorite. Geochim. Cosmochim. Acta **32**, 1013—1018 (1968).

— HESS, D. C.: Determination of very small quantities of lead. Anal. Chem. **32**, 960—966 (1960).

124 References

MASUDA, A.: A lead isotope composition in volcanic rocks of Japan. Geochim. Cosmochim. Acta **28**, 291—303 (1964).

MILLARD, H. T., JR.: Quantitative radiochemical procedure for analysis of polonium-210 and lead-212 in minerals. Anal. Chem. **35**, 1017—1023 (1963).

MINEEV, D. A.: Epidote containing rare earths from pegmatites of the Middle Urals. Akad. Nauk SSSR Doklady **127**, 865—868 (1959).

MIRKINA, S. L., GERLING, E. K., SHUKOLYUKOV, Y. A.: Determination of the absolute age of the alkaline complexes of the Middle Urals with the aid of the lead-isotopic and the potassium-argon methods. Geokhimiya, no. 8, 643—648 (1962).

— ISKANDEROVA, A. D.: The absolute age of some pegmatities of northern Karelia. Inform. Sb. Vses. Nauchno-Issled. Geol. Inst., no. 54, 117—126 (1962).

— MAKAROCHKIN, B. A.: On the use of minerals containing great amounts of common lead for the determination of the absolute age of post-protozoic formations (in Russian with English summ.). Geokhimiya, no. 8, 917—922 (1966).

MONTGOMERY, A.: Pre-Cambrian geology of the Picturis Range, north-central New Mexico. Bur. Mines and Mineral Res. Bull. **30**, 89 p. (1953).

MOORBATH, S.: Lead isotope abundance studies on mineral occurrences in the British Isles and their geological significance. Phil. Trans. Roy. Soc. London **254**, 295—360 (1962).

— WELKE, H.: Isotopic evidence for the continental affinity of the Rockall Bank, North Atlantic. Earth Planet. Sci. Letters **5**, 211—216 (1968a).

— — Lead isotope studies on igneous rocks from the Isle of Skye, northwest Scotland. Earth Planet. Sci. Letters **5**, 217—230 (1968b).

— — GALE, N. H.: The significance of lead isotope studies in ancient, high-grade metamorphic basement complexes, as exemplified by the Lewisian rocks of northwest Scotland. Earth Planet. Sci. Letters **6**, 245—256 (1969).

MUFFLER, L. J. P., DOE, B. R.: Composition and mean age of detritus of the Colorado River Delta in the Salton Trough, southeastern California. J. Sediment. Petrol. **38**, 384—399 (1968).

MURTHY, V. R.: Personal communication. In: MURTHY, V. R., PATTERSON, C. C.: Primary isochron of zero age for meteorites and the earth. J. Geophys. Res. **67**, (1962).

— PATTERSON, C. C.: Lead isotopes in ores and rocks of Butte, Montana. Econ. Geol. **56**, 59—67 (1961).

— — Primary isochron of zero age for meteorites and the earth. J. Geophys. Res. **67**, 1161—1167 (1962).

NARBUTT, K. I., LAPUTINA, I. P., SHUBA, I. D., KARDAKOV, K. A., SAMOILOV, G. P.: Isotopic composition of lead from ores, and age of minerals containing U, Th, and Pb, according to mass-spectrometric and X-ray spectrographic data. Akad. Nauk SSSR Inst. Geol. Rudnykh Mestorozhdeniy, Petrografii, Mineralogii i Geokhimii Trudy, no. 28, 122—137 (1959).

NAYDENOV, B. M., CHERDYNTSEV, V. V.: Change in the isotope composition of lead during separation from natural minerals. Akad. Nauk SSSR Izv. Ser. Geol. **5**, 40—49 (1958).

NICOLAYSEN, L. O.: Solid diffusion in radioactive minerals and the measurement of absolute age. Geochim. Cosmochim. Acta **11**, 41—59 (1957).

— BURGER, A. J., LIEBENBERG, W. R.: Evidence for the extreme age of certain minerals from the Dominion Reef conglomerates and the underlying granite in the Western Transvaal. Geochim. Cosmochim. Acta **26**, 15—23 (1962).

— DE VILLIERS, J. W. L., BURGER, A. J., STRELOW, F. W. E.: New measurements relating to the absolute age of the Transvaal system and of the Bushveld igneous complex (with discussion). Geol. Soc. South Africa Trans. **61**, 137—166 (1958).

NIER, A. O.: Variations in the relative abundances of the isotopes of common lead from various sources. Am. Chem. Soc. J. **60**, 1571—1576 (1938).

— The isotopic constitution of radiogenic leads and the measurement of geologic time, II. Phys. Rev. **55**, 153—163 (1939).

— THOMPSON, R. W., MURPHEY, B. F.: The isotopic constitution of lead and the measurement of geologic time, III. Phys. Rev. **60**, 112—116 (1941).

OOSTHUYZEN, E. J., BURGER, A. J.: Radiometric dating of intrusives associated with the Waterberg System. South Africa Geol. Survey Ann. **3**, pt. 2, 87—106 (1965).

OSTIC, R. G., RUSSELL, R. D., STANTON, R. L.: Additional measurements of the isotopic composition of lead from stratiform deposits. Can. J. Earth Sci. **4**, 245—269 (1967).

OVERSBY, V. M.: The isotopic composition of lead in iron meteorites. Geochim. Cosmochim. Acta **34**, 77—88 (1970).

— GAST, P. W.: Ocean basalt leads and the age of the earth. Science **162**, 925—927 (1968a).

— — Lead isotopic compositions and uranium decay series disequilibrium in recent volcanic rocks. Earth Planet. Sci. Letters **5**, 199—206 (1968b).

PASTEELS, P.: Age measurements on the zircons of some rocks from the Alps (in French, with English abs.). Schweiz. Mineral. Petrog. Mitt. **44**, 519—541 (1964).

— Investigations of the radioactive equilibrium in zircon by determination of the ratio $^{210}Pb/^{238}U$ (in French with English summ.). Cong. Soc. Savantes, Paris, Sec. Sci. Compt. Rend. **90**, pt. 2, 199—203 (1965).

PATTERSON, C. C.: The isotopic composition of meteoric, basaltic and oceanic leads, and the age of the earth [summary]. Conf. on Nuclear Processes in Geologic Settings, Proc. 36—40 (1953).

— The Pb^{207}/Pb^{206} ages of some stone meteorites: Geochim. Cosmochim. Acta **7**, 151—153 (1955).

— Age of meteorites and the earth. Geochim. Cosmochim. Acta **10**, 230—237 (1956).

— Characteristics of lead isotope evaluation on a continental scale in the earth, Chap. 19, pp. 244—268. In: Isotopic and Cosmic Chemistry, 553 p. Amsterdam: North-Holland Pub. Co. 1964.

— DUFFIELD, B.: The isotopic composition of lead in Easter Island rhyolite. Geochim. Cosmochim. Acta **27**, 1180—1181 (1963).

— TATSUMOTO, M.: The significance of lead isotopes in detrital feldspar with respect to chemical differentiation within the Earth's mantle. Geochim. Cosmochim. Acta **28**, 1—22 (1964).

— TILTON, G. R., INGHRAM, M. G.: Age of the earth. Science **121**, 69—75 (1955).

PEARSON, R. C., TWETO, O. L., STERN, T. W., THOMAS, H. H.: Age of Larimide porphyries near Leadville, Colorado. In: Short papers in geology and hydrology. U.S. Geol. Survey Prof. Paper 450-C, C78-C80 (1962).

PEIRSON, C. H., CAMBRAY, R. S., SPICER, G. S.: Lead-210 and polonium-210 in the atmosphere. Tellus **18**, 427—433 (1966).

PETERMAN, Z. E., DOE, B. R., BARTEL, A.: Data on the rock GSP-1 (granodiorite) and the isotope-dilution method of analysis for Rb and Sr. In: Geological Survey research 1967. U.S. Geol. Survey Prof. Paper 575-B, B181-B186 (1967).

— — PROSTKA, H. J.: Lead and strontium isotopes in rocks of the Absaroka volcanic field, Wyoming. Contr. Mineral. Petrol., in press (1970).

PICCIOTTO, E. E.: Measurement of the radioactivity of the air in the Antarctic. Nuovo Cimento 10, 190—191 (1958).

— WILGAIN, S.: Confirmation of the half-life of thorium-232 (in French). Nuovo Cimento 4, 1525—1528 (1956).

PIDGEON, R. T., O'NEIL, J. R., SILVER, L. T.: Uranium and lead isotopic stability in a metamict zircon under experimental hydrothermal conditions. Science 154, 1538—1540 (1966).

POLEVAYA, N. I., PANTELEEV, A. I.: Possibility of using the lead isotopic method for determining the age of glauconite (in Russian). Inform. Sb., Vses. Nauchno-Issled. Geol. Inst., no. 54, 31—36 (19

RAMTHUN, H.: New calorimetric determination of the half-life period of RaD (^{210}Pb). Z. Naturforsch. 19a, 1064—1069 (1964).

REED, G. W., KIGOSHI, K., TURKEVICH, A.: Determinations of concentrations of heavy elements in meteorites by activation analysis. Geochim. Cosmochim. Acta 20, 122—140 (1960).

REYNOLDS, P. H.: A lead isotope study of ores and adjacent rocks. Vancouver, British Columbia Univ., Ph. D. thesis, 94 p. (1967).

— RUSSELL, R. D.: Isotopic composition of lead from Balmat, New York. Can. J. Earth Sci. 5, 1239—1245 (1968).

ROBINSON, S. C., LOVERIDGE, W. D., RIMSAITE, J., VAN PETEGHEM, J.: Factors involved in discordant ages of euxenite from a Greenville pegmatite. Can. Mineral. 7, pt. 3, 533—546 (1963).

ROSE, H., JR., STERN, T. W.: Spectrochemical determination of lead in zircon for lead-alpha age measurements. Am. Mineral. 45, 1243—1256 (1960).

ROSHOLT, J. N., DOE, B. R., TATSUMOTO, M.: Evolution of the isotopic composition of uranium and thorium in soil profiles. Geol. Soc. Am. Bull. 77, 987—1004 (1966).

— PETERMAN, Z. E.: Uranium, thorium, lead systematics in the Granite Mountain, Wyoming. In: Abstracts with Programs for 1969, Pt. 5. Geol. Soc. America, p. 70 (1969).

— — BARTEL, A. J.: U-Th-Pb and Rb-Sr ages in granite reference from southwestern Saskatchewan. Canadian Jour. Earth Sci. 7, 184—187 (1970).

RUSSELL, R. D., FARQUHAR, R. M.: Lead isotopes in geology, 243 p. New York: Interscience Publ. 1960.

— KANASEWICH, E. R., OZARD, J. M.: Isotopic abundances of lead from a "frequently mixed" source. Earth Planet. Sci. Letters 1, 85—88 (1966).

— ULRYCH, T. J., KOLLAR, F.: Anomalous leads from Broken Hill, Australia. J. Geophys. Res. 66, 1495—1498 (1961).

RUTHERFORD, E.: Origin of actinium and age of the earth. Nature 123, 313—314 (1929).

SHIELDS, W. R. (Ed.): Analytical mass spectrometry section; instrumentation and procedures for isotopic analysis. Natl. Bur. Std. (U.S.) Tech. Notes, no. 277, 99 p. (1966).

— Analytical spectrometry section; summary of activities July 1966 to June 1967. Natl. Bur. Std. (U.S.) Tech. Notes, no. 426, 53 p. (1967).

SILVER, L. T.: The use of cogenetic uranium-lead isotope systems in zircons in geochronology. In: Radioactive dating. Internat. Atomic Energy Agency Symposium, Athens, 1962, Proc., 279—287 (1963).

— Uranium-thorium-lead isotope relations in some Lunar materials collected by Apollo 11. Science 167, 468—470 (1970).

— DEUTSCH, S.: Uranium-lead isotopic variations in zircons — A case study. J. Geol. 71, no. 6, 721—758 (1963).

SINCLAIR, A. J.: Anomalous lead from the Kootenary arc, British Columbia. In: A Symposium on the Tectonic History and Mineral Deposits of the Western Cordillera, Vancouver, B. C., 1964. Can. Inst. Min. and Metall. Spec. Vol. 8, 249—262 (1966).

SOBOTOVICH, E. V.: Possibility of determining the absolute age of the granites of the Terskey Ala-Tau by the lead included in them. Akad. Nauk SSSR Kom. Opredeleniyu Absolyut. Vozrasta Geol. Formatsii Trudy, sess. 9, 269—280 (1961).

— GRASHCHENKO, S. M.: Isotopic composition of recent lead as age criterion for isolated igneous rock samples (in Russian). Akad. Nauk SSSR Izv. Ser. Geol. 30, 3—9 (1965).

— — LOVTSYUS, A. V.: Rock age of Taromskoe quarry according to data of lead isochronous method. Akad. Nauk SSSR Kom. Opredeleniyu Absolyut. Vozrasta Geol. Formatsii Trudy, sess. 11, 353—356 (1963a).

— — — Isotopic composition of lead from very old rocks. Radiokhimiya 5, 157—160 (1963b).

— — ALEKSANDRUK, V. M., SHATS, M. M.: Determination of the age of the most ancient rocks by the lead isochronous and the strontium isotope-spectral methods. Akad. Nauk SSSR Izv. Ser. Geol. 28, 3—14 (1963c).

— — LOVTSYUS, A. V.: Age of the rocks of the Sharyzhalgay series (Baikal block). Akad. Nauk SSSR Izv. Ser. Geol., no. 9, 38—41 (1965).

— LOVTSYUS, G. P., LOVTSYUS, A. V.: New data on the content and isotopic composition of lead in stony meteorites. Akad. Nauk SSSR Meteritika, no. 24, 29—33 (1964).

SOMAYAJULU, B. L. K., TATSUMOTO, M., ROSHOLT, J. N., KNIGHT, R. J.: Disequilibrium of the ^{238}U series in basalt. Earth Planet. Sci. Letters 1, 387—391 (1966).

STACEY, J. S., DELEVAUX, M. E., ULRYCH, T. J.: Some triple-filament lead isotope ratio measurements and an absolute growth curve for single-stage leads. Earth Planet. Sci. Letters 6, 16—25 (1969).

— ZARTMAN, R. E., NKOMO, I. T.: A lead isotope study of galenas and selected feldspars from mining districts in Utah. Econ. Geol. 63, 796—814 (1968).

STANTON, R. L., RUSSELL, R. D.: Anomalous leads and the emplacement of lead sulfide ores. Econ. Geol. 54, 588—607 (1959).

STARIK, I. E., LAZAREV, K. F.: Study of the comparative leachability of the isotopes of radium, uranium, and thorium from monazite. Methody Opredeleniyu Absolyut. Vozrasta Geol. Obrazov., no. 6, 24—31 (1964).

— LOVTSYUS, G. P., SOBOTOVICH, E. V., GRASHCHENKO, S. M., SHATS, M. M., LOVTSYUS, A. V.: Isotopic composition of lead in meteorites and the problems of their origin. Akad. Nauk SSSR Kom. Opredeleniyu Absolyut. Vozrasta Geol. Formatsii Byull., no. 5, 12—25 (1962a).

— SHATS, M. M., SOBOTOVICH, E. V.: On the age of meteorites. Akad. Nauk SSSR Doklady 123, 424—426 (1958a).

— SOBOTOVICH, E. V.: Determination of the isotope composition of lead in rocks. Akad. Nauk SSSR Doklady 111, 395 (1956).

— — AVDZEIKO, G. V., LOVTSYUS, A. V.: A new method of determination of the isotopic composition of leads in rocks. Akad. Nauk SSSR Kom. Opredeleniyu Absolyut. Vozrasta Geol. Formatsii Trudy, 5th sess., 233—242 (1958b).

— — LOVTSYUS, G. P.: The heterogeneity of lead in natural formations. Akad. Nauk SSSR Kom. Opredeleniyu Absolyut. Vozrasta Geol. Formatsii Byull., no. 3, 54—59 (1958c).

— — — SHATS, M. M., LOVTSYUS, A. V.: On the problem of the isotopic composition of lead of iron meteorites. Akad. Nauk SSSR Meteoritika, no. 20, 103—113 (1961).

— — SHATS, M. M.: The problem of the origin of meteorites and tektites. Geo-khimiya, no. 3, 245—251 (1963).

— STARIK, F. E., PETRYAYEV, E. P., LAZAREV, K. F., YELIZAROVA, A. N.: Sig-nificance of migration of radioactive elements from minerals for the deter-mination of their age by the lead method. Internat. Geol. Cong. 21st, Copen-hagen, 1960, Doklady Sovet. Geol. Problema 3, 15—31 (1960).

— — YELIZAROVA, A. N.: Comparative leachability of some isotopes. Akad. Nauk SSSR Kom. Opredeleniyu Absolyut. Vozrasta Geol. Formatsii Byull., no. 4, 160—165 (1961).

— — — PETRYAYEV, E. P., LAZAREV, K. F.: The significance of diffusion of various radioactive elements in age determination by the lead method. Akad. Nauk SSSR Kom. Opredeleniyu Absolyut. Vozrasta Geol. Formatsii Trudy, 5th sess., 221—232 (1958d).

— VOROB'YEV, G. G., SOBOTOVICH, E. V., SHATS, M. M., GRASHCHENKO, S. M.: Origin and age of tektites. Akad. Nauk SSSR Kom. Opredeleniyu Absolyut. Vozrasta Geol. Formatsii Byull., no. 5, 26—34 (1962b).

STEIGER, R. H., WASSERBURG, G. J.: Systematics in the Pb^{208}-Th^{232}, Pb^{207}-U^{235}, and Pb^{206}-U^{238} systems. J. Geophys. Res. 71, 6065—6090 (1966).

— — Effects of acid-washing procedure on Th-Pb ages of zircons. In: Abstracts for 1967. Geol. Soc. Am. Spec. Paper 115, 213 (1968).

— — Comparative U-Th-Pb systematics in 2.7×10^9 yr plutons of different geologic histories. Geochim. Cosmochim. Acta 33, 1213—1232 (1969).

STERN, T. W., GOLDICH, S. S., NEWELL, M. F.: Effects of weathering on the U-Pb ages of zircon from the Morton Gneiss, Minnesota. Earth Planet. Sci. Letters 1, 369—371 (1966).

— NEWELL, M. F., KISTLER, R. W., SHAWE, D. R.: Zircon uranium-lead and thorium-lead ages and mineral potassium-argon age of La Sal Mountains rocks, Utah. J. Geophys. Res. 70, 1503—1507 (1965).

— ROSE, H. J., JR.: New results from lead-alpha age measurements. Am. Mineral. 46, 606—612 (1961).

STIEFF, L. R., STERN, T. W.: Graphic and algebraic solutions of the discordant lead-uranium age problem. Geochim. Cosmochim. Acta 22, 176—199 (1961).

— — EICHER, R. N.: Algebraic and graphic methods for evaluating discordant lead-isotopes ages. U.S. Geol. Survey Prof. Paper 414-E, E1-E27 (1963).

STRELOW, F. W. E., TOERIEN, F. VON S.: Separation of lead (II) from bismuth (III), thallium (III), cadmium (II), mercury (II), gold (III), platinum (IV), palladium (II), and other elements by anion exchange chromatography. Anal. Chem. 38, 545—548 (1966).

TATSUMOTO, M.: Isotopic composition of lead in volcanic rocks from Hawaii, Iwo Jima, and Japan. J. Geophys. Res. 71, 1721—1733 (1966a).

— Genetic relations of oceanic basalts as indicated by lead isotopes. Science 153, 1094—1101 (1966b).

— Lead isotopes in volcanic rocks and possible ocean-floor thrusting beneath island arcs. Earth Planet. Sci. Letters **6**, 369—376 (1969).

— KNIGHT, R. J.: Isotopic composition of lead in volcanic rocks from central Honshu — with regard to basalt genesis. Geochem. J. **3**, 53—86 (1969).

— PATTERSON, C. C.: The concentration of common lead in sea water, pp. 74—89. In: Earth Science and Meteorics, 312 p. Amsterdam: North-Holland Publ. Co. 1963.

— ROSHOLT, J. N.: The age of the moon. Science **167**, 461—462 (1970).

— SNAVELY, P. D., JR.: Isotopic composition of lead in rocks of the Coast Range, Oregon and Washington: J. Geophys. Res. **74**, 1087—1100 (1969).

TILTON, G. R.: Isotopic composition of lead from tektites. Geochim. Cosmochim. Acta **14**, 323—330 (1958).

— Volume diffusion as a mechanism for discordant lead ages. J. Geophys. Res. **65**, 2933—2945 (1960).

— GRÜNENFELDER, M. H.: Sphene, uranium-lead ages. Science **159**, 1458—1461 (1968).

— NICOLAYSEN, L. O.: The use of monazite for age determination. Geochim. Cosmochim. Acta **11**, 28—40 (1957).

— PATTERSON, C. C., BROWN, H., INGHRAM, M., HAYDEN, R., HESS, D., LARSEN, E., JR.: Isotopic composition and distribution of lead, uranium, and thorium in a Precambrian granite (Ontario). Geol. Soc. Am. Bull. **66**, 1131—1148 (1955).

— STEIGER, R. H.: Lead isotopes and the age of the Earth. Science **150**, 1805—1808 (1965).

— — Mineral ages and isotopic composition of primary lead at Manitouwadge, Ontario. J. Geophys. Res. **74**, 2118—2132 (1969).

TRIPPLER, K.: Report on investigations of the β-activity of the atmosphere near the ground. Z. Geophysik **32**, 102—112 (1966).

TUGARINOV, A. I.: The definition of the time of metamorphism of altered uranium minerals and measurement of absolute age. Akad. Nauk SSSR Kom. Opredeleniyu Absolyut. Vozrasta Geol. Formatsii Byull., no. 2, 82—89 (1957).

— GAVRILOVA, L. K., BEDRINOV, V. P.: Evolution of the isotopic composition of lead in Precambrian granitic rocks. Vopr. Prikladn. Radiogeol. 228—243 (1963a).

— KOVALENKO, V. I., ZNAMENSKIY, YE. B., LEGEYDO, V. A., SOBOTOVICH, E. V., BRANDT, S. B., TSYKHANSKIY, V. D.: Distribution of Pb isotopes, Sn, Nb, Ta, Zr, and Hf ingranitoids of Nigeria, pp. 687—699. In: Origin and Distribution of the Elements (AHRENS, L. A., ed.). Oxford: Pergamon Press 1968.

— VOYTKEVICH, G. V.: Precambrian geochronology of the continents. Moscow, Izdatel'stvo "Nedra", 388 p. (1966).

— ZYKOV, S. I., BIBIKOVA, E. V.: On the determination of the absolute age of sedimentary rocks by the lead-uranium method. Geokhimiya, no. 3, 266—283 (1963b).

ULRYCH, T. J.: Oceanic basaltic leads — A new interpretation and an independent age for the Earth. Science **158**, 252—256 (1967).

— BURGER, A., NICOLAYSEN, L. O.: Least radiogenic terrestrial leads. Earth. Planet. Sci. Letters **2**, 179—184 (1967).

— RUSSELL, R. D.: Gas source mass spectrometry of trace leads from Sudbury, Ontario. Geochim. Cosmochim. Acta **28**, 455—469 (1964).

VENKATASUBRAMANIAN, V. S.: Studies on radon leakage in minerals. In: Geophysical Exploration, A Symposium, Baroda, India, 1959, Proc. New Delhi, Council Sci. Indust. Res., 102—105 (1963).

VILENSKII, V. D., DAVYDOV, E. N., MALAKHOV, S. G.: The seasonal and geographical changes of lead-210 content in the atmosphere. Radioaktivn. Izotopy v Atm. i ikh Ispol'z. v Meteorol., Nauchno. Konf. Yadern. Meteorol., Obninsk, USSR, 1964, 120—131 (1965).

VINOGRADOV, A. P.: Comparison of data on the age of rocks obtained by different methods and geological conclusions. Geokhimiya, no. 5, 3—17 (1956).

— ZADOROZHNYI, I. K., ZYKOV, S. I.: Isotopic composition of lead and the age the earth. Akad. Nauk SSSR Doklady 85, 1107—1110 (1952).

— ZYKOV, S. I.: New data on the lead isotope composition. Akad. Nauk SSSR Doklady 105, 126—128 (1955).

VOLOVYEV, M. I., ZYKOV, S. I., STUPNIKOVA, N. I., MUSATOV, D. I., ZATSEPINA, E. F.: Interpretation of absolute age values of rock forming accessory minerals in the Enisei Ridge and Eastern Sayan Mountains (in Russian). Novye Dannye Geol. Yuga Krasnoyar. Kraya, Krasnoyarsk, Sb., 272—294 (1963).

WAMPLER, J. M., KULP, J. L.: Isotopic composition and concentration of lead in some carbonate rocks. In: Petrologic Studies, pp. 105—114. Geol. Soc. Am. *Buddington Vol.*, 660 p. (1962).

— — An isotopic study of lead in sedimentary pyrite. Geochim. Cosmochim. Acta 28, 1419—1458 (1964).

— SMITH, D. H., CAMERON, A. E.: Isotopic comparison of lead in tektites with lead in earth materials. Geochim. Cosmochim. Acta 33, 1045—1055 (1969).

WASSERBURG, G. J.: Diffusion processes in lead-uranium systems. J. Geophys. Res. 68, 4823—4846 (1963).

— Geochronology and isotopic data bearing on development of the continental crust. Adv. Earth Sci., Contrib. Internat. Conf., Cambridge, Mass., 1964, 431—459 (1966).

— HAYDEN, R. J.: A^{40}-K^{40} dating. Geochim. Cosmochim. Acta 7, 51—60 (1955).

— TOWELL, D., STEIGER, R. H.: A study of Rb-Sr systematics in some Precambrian granites of New Mexico [abs.]: Am. Geophys. Union Trans. 46, 173—174 (1965).

WEDEPOHL, K. H.: Investigation of the Kupferschiefer in northwest Germany; a contribution to the explanation of the genesis of bituminous sediments. Geochim. Cosmochim. Acta 28, 305—364 (1964).

WELIN, E., BLOMQVIST, G.: Age measurements on radioactive minerals from Sweden. Geol. Foren. Stockholm Forh. 86, 33—50 (1964).

WELKE, H., MOORBATH, S., CUMMING, G. L., SIGURDSSON, H.: Lead isotope studies on igneous rocks from Iceland. Earth Planet. Sci. Letters 4, 221—231 (1968).

WETHERILL, G. W.: An interpretation of the Rhodesia and Witwatersrand age patterns. Geochim. Cosmochim. Acta 9, 290—292 (1956a).

— Discordant uranium-lead ages. Am. Geophys. Union Trans. 37, 320—326 (1956b).

— Discordant uranium-lead ages — [Pt.] 2, Discordant ages resulting from diffusion of lead and uranium. J. Geophys. Res. 68, 2957—2965 (1963).

— WASSERBURG, G. J., ALDRICH, L. T., TILTON, G. R., HAYDEN, R. J.: Decay constants of K^{40} as determined by the radiogenic argon content of potassium minerals. Phys. Rev. 103, 987—989 (1956).

YEATS, R. S.: Southern California structure, sea-floor spreading, and history of the Pacific Basin. Geol. Soc. Am. Bull. 79, 1693—1702 (1968).

ZARTMAN, R. E.: The isotopic composition of lead in microlines from the Llano uplift, Texas. J. Geophys. Res. 70, 965—975 (1965).

— Lead isotopes in igneous rocks of the Grenville Province as a possible clue to the presence of colder crust. Geol. Assoc. Can. Spec. Paper 5, 193—205 (1969).

— WASSERBURG, G. J.: The isotopic composition of lead in potassium feldspars from some 1.0-b.y.-old North American igneous rocks. Geochim. Cosmochim. Acta **33**, 901—942 (1969).

ZHIROV, K. K., ZYKOV, S. I.: Metamorphism and the time of formation of granites as based on isotopic analyses of lead. Geochem., no. 7, 684—695 (1956).

— — ZHIROVA, V. V., STUPNIKOVA, N. I.: Effects of hydrothermal alteration upon calculation of ages of radioactive minerals. Geokhimiya, no. 8, 657—665 (1957).

ZHIROVA, V. V., ZYKOV, S. I., TUGARINOV, A. I.: On the age of zircons of most ancient formations of the Kola peninsula. Geokhimiya, no. 12, 1043—1052 (1961).

ZYKOV, S. I., TUGARINOV, A. I., BEL'KOV, I. V., BIBIKOVA, E. V.: The age of the most ancient formations of the Kola Peninsula. Geokhimiya, no. 4, 307—314 (1964).

Subject Index

Minerals, Rocks and Inorganic Materials
